Study Guide for

Chemistry

by

Steven S. Zumdahl

Martha B. Barrett

University of Colorado at Denver

D. C. Heath and Company

Lexington, Massachusetts   Toronto

Cover photo: Camera graphics produced on an MC 1600 computerized optical printer. William Smyth/Slide Graphics of New England.

Copyright © 1986 by D. C. Heath and Company.

Published simultaneously in Canada.

Printed in the United States of America.

International Standard Book Number: 0-669-04554-3

TABLE OF CONTENTS

INTRODUCTION

The study of chemistry offers many challenges to the beginning student.

- Chemistry cannot be presented in a "linear" fashion. In some respects, we must leap into the center of the subject, keep building on what we've studied, but keep returning to expand topics we have already covered.

- Much of chemistry is not "concrete," but requires abstract models. When you study introductory physics you can experiment with blocks sliding on planes, or balls travelling through trajectories. But you cannot hold an atom or molecule in your hand, or watch an electron in motion. When you are using a model, remember that the model is not the reality, but a simplified way of picturing a reality.

- You may not feel comfortable with chemistry during your introductory course: there are many topics to cover, and the pace is often swift. Some students read the text over and over, and the study guide as well, waiting for the subject to become comfortable before they attack the problems. Remember, chemistry is not a subject that can be mastered by reading. In spite of your discomfort, plunge into the problems: much of learning chemistry comes by doing.

Hints for Studying Chemistry

- Don't get behind! At most colleges and universities, at least three hours of work are expected outside of class each week for every hour in class. Don't expect to master the material in one long study session each week. Study chemistry often in shorter sessions.

- Read the assigned material in the text before it is covered in lecture: scan the chapter, pick out the key words, look at the sample problems. Then read the material in earnest; work out the sample problems. Don't expect the material to be crystal clear on the first pass. Do pick out the concepts that seem unclear. Try some of the basic problems at the end of the chapter.

- Take notes in lecture, but not at the expense of listening to, and trying to understand the lecture. Mark the material you don't understand in your notes.

- Ask questions! The only "dumb" question is the one you didn't ask. Some instructors encourage questions in class. If it is not appropriate to ask questions during a lecture, be sure to see your instructors after class, in the lab, or during office hours. Don't wait! New material lurks in the next lecture.

- Read the material in the text again. Do the problems.

- Don't spend too much time on any one problem. If you haven't made a good start within about 20 minutes, move on to another problem, or get help from an instructor or classmate.

- Do as many problems as you can.

- Make a plan for your solution before you begin; the study guide will give you examples.

- Write out the solutions to your problems: make the method of solution clear; this will help you when you review, and is good practice for your exams and quizzes.

- Carry out your solutions with appropriate units and significant figures.

- Does the numerical answer make sense? Do the units on the answer make sense? If not, go back and look for errors.

- Don't abuse the solutions, if they are available. It is too easy to look at someone else's solution to say, "Oh yes, I understand how to do that." The challenge is to be able to solve the problems yourself. On the other hand, if you are making absolutely no headway on a problem after a solid effort, reach for the solutions. Once that problem is clear, solve another one like it, to check your understanding.

- Don't memorize methods of solving problems. Be able to think out the solution. This study guide will give you some hints.

- If you are using a calculator on your exams, use your calculator to solve all your problems. Be comfortable with your calculator before your exams.

- Find a "buddy" or two to study with. Sometimes your understanding of a topic is improved by explaining it to someone else. Sometimes your fellow students understand your difficulties better than your instructor. Remember, though, in the end, it is your understanding that matters. Don't let your buddies do all the work.

How to Use This Study Guide

Your text has lots of sample problems, with solutions, a list of key terms and a chapter summary. Each section has a purpose listed, giving you your objectives for the section. The study guide will focus on areas where some students need extra help. The guide provides more examples, different ways of solving problems, different approaches to key concepts. Each chapter of the guide provides a quick short answer self-test. This self-test will help you pick out areas that you need to concentrate on.

Studying for an Exam

- Go over your notes and the problems that you've done.

- Do more problems for practice; do the problems with your book closed.

- Design your own practice test. What kinds of problems did your instructor focus on in class? Instructors all have their own styles--each will give a different sort of exam. What do you know about your instructor's style? What kinds of problems were assigned? What sorts of questions can you design to highlight the concepts you have covered? Remember, there will be little rote memorization; the focus will be on understanding concepts and solving problems.

- Be sure to get enough sleep the night before the exam. Make sure that your calculator batteries are charged, and that you have sharpened pencils ready.

Taking the Exam

- Take a deep breath. Relax. A little adrenalin improves your performance. Too much adrenalin can paralyze you. Remember, you've done lots of problems. And, your chemistry exam is designed to test your mastery of chemistry, not your unique value as a human being.

- Scan the entire exam before you begin.

- Pace yourself: if your exam is worth 100 points, and you have 50 minutes, expect to complete at least 10 points of questions in 5 minutes. Don't spend too long on any one problem.

- Do the problems or questions you find easiest first: remember, most exam questions are not arranged in order of difficulty. You may find an easy question at the end of the exam.

- Do set the problem up, before you carry out any calculations. Make your method of solution clear.

- Carry out your calculations with significant figures and appropriate units.

- Check that your answer makes sense. (These steps are ones you've practiced during your problem solving. By now they should feel comfortable.)

After the Exam

- While the exam is fresh in your mind, look at your text and your notes to clear up some of the puzzles. Then put thoughts of the exam aside.

- Your instructor may go over the exam in class, or post solutions to the exam. Figure out what mistakes you made, and, if you can, why you made them. What can you do differently next time? What did you learn about the kinds of exams your instructor gives? Remember, your study of chemistry keeps building on the material you've already covered. If there are holes in your understanding, they will plague you again, either in your study of new material, or on your final exam.

Using Your Calculator

Take time to explore the instruction book for your calculator. Then practice. Use your calculator to solve the problems in the text. Use your calculator to perform calculations in the laboratory. You need to be comfortable with your calculator before you face your first exam.

Entering Data

There are two different methods of entering data into calculators: RPN (Reverse Polish Notation) and algebraic notation (direct formula entry).

RPN resembles computer programming. Calculators using RPN have an ENTER key. To add 5 and 7 on an RPN calculator the key sequence is: 5   ENTER   7   + ; to multiply 5 and 7 the key sequence is: 5   ENTER   7   × .

You can enter a formula directly as written into a calculator using algebraic notation. To add 5 and 7 on a calculator using algebraic notation the key sequence is: 5   +   7   = ; to multiply 5 and 7 the key sequence is: 5   ×   7   = . Calculators using algebraic notation have an equals (=) key.

Operations

You can carry out a series of operations on your calculator. But, be aware that each calculator has its own priorities for carrying out operations in a string.

If you have a calculator using algebraic notation try this key sequence.

$$2 \ + \ 3 \ \times \ 3 \ =$$

What is your answer? 11. 2 + 3 × 3 is translated by your calculator into 2 + (3 × 3). Your calculator completes multiplication before it completes addition. If you wanted to calculate (2 + 3) × 3 you would use parentheses [(2 + 3) × 3 =] or use the "equals" key after 2 + 3: [2 + 3 = × 3 =].

Exercise care when "stringing together" division and multiplication. For example, to calculate

$$\frac{1.45 \times 2.62}{17.1 \times .543} =$$

Multiply 1.45 by 2.62, divide by 17.1 and divide by .543.

Estimate answers to catch errors. For example, in the preceding calculation we could estimate that the answer would be approximately

$$\frac{1.5}{15} \times \frac{2.5}{.5} \approx \frac{1}{10} \times 5 \approx \frac{1}{2} = .500$$

Check the answer on your calculator. [.409141331...]. If you found .1206... you made an error in your calculations. Can you find the error?

Significant Figures

A calculator may show too many significant figures after a calculation. We must round off the answer above to three significant figures, .409. Sometimes a calculator shows too few significant figures, Calculate

$$\frac{2556}{1278}$$

Some calculators do not show "trailing" zeros unless they are needed to locate a decimal point; these calculators show

$$\frac{2556}{1278} = 2,$$

not 2.000. Remember, you need to keep track of significant figures: your calculator won't.

Scientific Notation

Learn to enter numbers in scientific notation.

To enter 1.5×10^6, use the key sequence

$$1 \quad \cdot \quad 5 \quad \text{EXP} \quad (\text{or EE↓}) \quad 6 \; .$$

Do not use the $\times$ key!

Your calculator reads 1.5 06.

To enter 7.2×10^{-5}, use the key sequence

$$7 \quad \cdot \quad 2 \quad \text{EXP} \quad +/- \quad (\text{or CHS}) \quad 5$$

Your calculator reads 7.2 − 05. If you use the "change sign" key (+/− or CHS) before you use the exponent key, you change the sign of the number; if you use the "change sign" key after you use the exponent key you only change the sign of the exponent.

Other Function Keys

Learn to:

Take logarithms to the base 10: LOG

Take logarithms to the base e: LN

Raise 10 to a power: 10^x or   INV   LOG

Raise a number y to a power x:

For example 3^4:   3 y^x 4   =

5

To take squares and square roots: x^2 , $\sqrt{x}$

To invert a number: $1/x$

Making Corrections

Learn to use your "clear" key [C , C•CE], to make corrections. Usually an incorrect entry can be replaced so long as it has not been followed by an operation. For example, if you enter   7 × 5 , on a calculator using algebraic notation but the 5 should be a 6, use your "clear" key once: C   6   =   42.   If the "clear" key is pressed twice, the calculator will be completely cleared.   If the "clear" key is pressed once after an operations key [×, = , etc.] the calculator will be completely cleared.

Units

Remember to include units in your answers – your calculator won't keep track of them for you.

Care and Feeding

Treat your calculator gently. Don't tuck it in your back pocket, as you might damage it when you sit down. Don't let your calculator be crushed by books in your back pack. Keep your calculator dry – avoid contact with coffee, soft drinks and wet bathing suits! Keep your calculator away from heat – avoid hot radiators in the winter and hot vehicles in the summer sun.

Use your calculator often. Return to your instruction book now and then to explore and learn more about your calculator.

If your calculator uses batteries you might want to replace them before they fail completely. Avoid having your calculator "die" during an exam.

Enjoy your calculator. It may be frustrating at times – but calculators are easier to use (and friendlier) than slide rules, log tables and/or hand calculation.

CHAPTER ONE: CHEMICAL FOUNDATIONS

Chemistry is an experimental science, a systematic study of matter, the "stuff" of the universe, and the changes that matter undergoes. Much of chemistry involves quantitative observations, measurements, and the search for patterns among the observations.

Each measurement must be recorded with <u>units</u> and some indication of the uncertainty of the measurement.

Most scientists, and most of the nonscientific world, use the metric system (SI) of units. Most Americans resist the metric system, wary of converting inches to centimeters (1 in = 2.54 cm), quarts to liters (1 qt = .9464 L), miles to kilometers (1 mi = 1.61 km). However, converting among units in the metric system is much easier than in the English system. And once we abandon the English system, these curious conversion factors will be of only historical interest.

You can begin to develop a "metric feel" for the world by remembering:

A dime is about 1 millimeter thick.

Your little finger is about 1 centimeter wide.

Your pace is about 1 meter long (a bit longer than 1 yard).

A cup of water is nearly 1/4 of a liter (.237 L) and weighs nearly 1/4 of a kilogram.

You will probably be asked to convert measurements between the English system and the metric system for practice, but most of a chemist's calculations stay within the metric system.

Many measurements in science deal with very small numbers (e.g., the mass of an atom), and very large numbers (e.g., the number of water molecules in a "teaspoonful"). To simplify writing these numbers, we'll use "scientific" (or "exponential" notation).

Exponents

First, a quick review of exponents: (See Appendix 1.1.)

Exponent

$$10^4 = 10 \times 10 \times 10 \times 10$$

Base

i) When multiplying two numbers expressed to the same base, add exponents:

$$10^4 \times 10^3 = (10 \times 10 \times 10 \times 10) \times (10 \times 10 \times 10) = 10^{4 + 3} = 10^7$$

ii) When dividing two numbers expressed to the same base, subtract exponents:

$$\frac{10^5}{10^2} = \frac{10 \times 10 \times 10 \times 10 \times 10}{10 \times 10} = 10^{5-2} = 10^3$$

Then,

$$\frac{10^2}{10^5} = \frac{10 \times 10}{10 \times 10 \times 10 \times 10 \times 10} = \frac{1}{10 \times 10 \times 10} = 10^{2-5} = 10^{-3}$$

Therefore, negative exponents represent numbers smaller than 1.

And: $\frac{10^3}{10^3} = 10^{3-3} = 10^0$, which is 1.

Problem 1-1:

Multiply: $(3.2 \times 10^4) \times (1.6 \times 10^2)$:

Solution:

Rearrange and multiply: $(3.2 \times 1.6) \times (10^4 \times 10^2) = 5.12 \times 10^{(4+2)}$

$$= 5.12 \times 10^6$$

Problem 1-2:

Divide: $\frac{3.51 \times 10^2}{1.56 \times 10^3}$

Solution:

Rearrange: $\frac{3.51}{1.56} \times \frac{10^2}{10^3}$

$$2.25 \times 10^{(2-3)} = 2.25 \times 10^{-1}$$

Exponential Notation

Any number can be written as a number between 1 and 10, times a power of ten. For example: 357 can be written as 3.57×10^2; .0273 can be written as 2.73×10^{-2}.

Problem 1-3:

   a.   Write 5081 in exponential notation:

Solution:

   Write 5081×10^0 (Remember $10^0 = 1$)

   Move the decimal point to the <u>left</u> three places, to give a number between 1 and 10.

   <u>Increase</u> the power of ten by 1 for each place you move the decimal point to the left.

$$5081. \times 10^0 = 5.081 \times 10^3$$

   b.   Write .0000769 in exponential form

Solution:

   Write $.0000769 \times 10^0$

   Move the decimal point to the <u>right</u> five places.

   <u>Decrease</u> the power of ten by 1 for each place you move the decimal point to the right.

$$.0000769 \times 10^0 = 7.69 \times 10^{-5}$$

Learn to enter exponential numbers on your calculator:

Check your instruction book:

For example, to enter 5.081×10^3 key in:

   5 . 0 8 1 EXP 3   (or 10^x : Find your exponent key)

   Your display will show 5.081   03

                  the coefficient   the power of 10

To enter: 7.69×10^{-5} key in:

   7 . 6 9 EXP +/- 5

   (Hint: Use the "Change Sign" key after the exponent key.)

   Your display should read 7.69   −05

                  the coefficient   the power of 10

9

Use your calculator to multiply:

$$(5.081 \times 10^{+3}) \times (7.69 \times 10^{-5}) = 3.907\ldots \times 10^{-1}$$

$$\text{or } .3907\ldots$$

Learn to estimate your answer:

$$5.081 \times 10^3 \text{ is about } 5 \times 10^3$$

$$7.69 \times 10^{-5} \text{ is about } 8 \times 10^{-5}$$

Our answer should be about $5 \times 8 \times 10^3 \times 10^{-5}$

$$\approx 40 \times 10^{-2}$$

$$\text{or } 4 \times 10^{-1} \text{ or } .4$$

You can use your estimated answer to check your calculator's answer (or your entry on the calculator).

The Metric System

The metric system uses prefixes to designate larger and smaller units of measure. For example, the prefix *centi* indicates 1/100th; a centimeter is 1/100th of a meter. The prefix *kilo* indicates 1000; a kilometer is 1000 meters. We can "convert" from one unit of measure to another by moving the decimal point the appropriate number of places.

Problem 1-4:

    a. A rod is 2.56 m (meters) long. How long is this rod in cm (centimeters)?

Solution:

1 meter = 100 centimeters

This rod is $2.56 \text{ m} \times \dfrac{100 \text{ cm}}{1 \text{ m}} = 256 \text{ cm}$

    b. A package weighs 468 grams. How many kilograms does it weigh?

Solution:

1 kilogram = 1000 grams

or 1 gram = $\dfrac{1}{1000}$ kilogram or $\dfrac{1 \text{ kilogram}}{1000 \text{ grams}}$

This package weighs: $468 \text{ g} \times \dfrac{1 \text{ kg}}{1000 \text{ g}} = .468 \text{ kg}$

Memorize the meaning of the most common prefixes in Table 1.2.

Uncertainty

Every measurement involves some uncertainty. If you measure the length of a book with a centimeter ruler, your result will depend on how you position the "zero" point of your ruler, the angle that you read the scale on the ruler, how you estimate the position of the edge of the book, how finely divided the scale is, and how accurate the divisions are.

You should estimate and record the uncertainty of every measurement you make. Record all the certain digits and the first uncertain (estimated) digit in your measurement. These digits are called "significant figures."

If you measure a book with your centimeter ruler, and you record 23.6 cm, you are indicating that the book is 23.6 ± .1 cm (or between 23.5 and 23.7 cm) long. If you record 23, you are indicating the book is between 22 and 24 cm long. 23.6 has three significant figures, 23 has two significant figures.

Often you will calculate a result by combining uncertain measurements. Each result will also be uncertain.

Problem 1-5:

For example, you can determine the density of an object in the lab by weighing it, and measuring the volume of water it displaces in a graduated cylinder.

If you record the following data:

Mass of the object: 16.721 g

Volume reading of water in the cylinder: 22.7 cm^3

Volume reading of water + object in the cylinder: 34.5 cm^3

a. What is the volume of the object?

Solution:

$$
\begin{array}{ll}
34.5 \text{ cm}^3 & \pm \ .1 \text{ cm}^3 \\
-22.7 \text{ cm}^3 & \pm \ .1 \text{ cm}^3 \\
\hline
11.8 \text{ cm}^3 & \text{The uncertainty in the volume is } \underline{\text{at least}} \pm \ .1 \text{ cm}^3
\end{array}
$$

b. What is the density?

Solution:

$$ d = \frac{m}{v} = \frac{16.721 \text{ g} \pm .001 \text{ g}}{11.8 \text{ cm}^3 \pm .1 \text{ cm}^3} $$

The calculator reads = 1.417033898 g/cm^3

How well do we __really__ know the density?

If our estimates of uncertainty are right:

The density could be as large as $\dfrac{16.721 + .001}{11.8 - .1} = 1.42923\ldots\ \dfrac{g}{cm^3}$

(using the largest mass and the smallest volume).

Or as small as $\dfrac{16.721 - .001}{11.8 + .1} = 1.405\ldots\ \dfrac{g}{cm^3}$

(using the smallest mass and the largest volume).

Therefore we should express the density as $\dfrac{1.42\ g}{cm^3}$.

$\left(\text{The uncertainty is about } \dfrac{.01\ g}{cm^3}\right)$.

Significant Figures

We'll use rules for "significant" figures to estimate the uncertainty in our calculated results.

1. In multiplication or division, the number of significant figures in a result is __the same__ as the number of significant figures in the input factor with the fewest number of significant figures.

   In our density problem, 16.721 has 5 significant figures;

   11.8 has 3 significant figures.

   Therefore, the density should have 3 significant figures

   $$1.417\ldots\ g/cm^3$$

   is closer to 1.42 than to 1.41.

   Thus our answer was 1.42 g/cm³.

   (Look at the "rounding rules" to come.)

2. In addition or subtraction, align the decimal points for the numbers and carry out the calculation. Then, find the first column from the left with an uncertain digit: that column determines the uncertain digit in your answer.

For example: Add 18.172, 5.18 and .0015

$$
\begin{array}{r}
18.172 \\
5.18 \\
\underline{.0015} \\
23.2535
\end{array}
$$

certain   first uncertain

The answer will be 23.25, with four significant figures.

Don't be misled by the enticing results on your calculator. Express each result with appropriate significant figures and units.

Hints: All non-zero digits are significant.

Leading zeros just determine the position of the decimal point and are not significant.

Trailing zeros after a decimal point or followed by a decimal point are significant.

Trailing zeros without a decimal point are ambiguous: these zeros may or may not be significant; you may have to decide from the context of the problem.

Numbers written in exponential form are not ambiguous.

Consider these numbers:

The leading zeros are not significant

$.00245 = 2.45 \times 10^{-3}$ 3 significant figures

$715.0 = 7.150 \times 10^{2}$ 4 significant figures

$5500 = 5.5 \times 10^{3}$ 2 significant figures

or $= 5.50 \times 10^{3}$ 3 significant figures

or $= 5.500 \times 10^{3}$ 4 significant figures

Some measured numbers are exact. Counted numbers are exact (e.g., 35 students, 144 pens).

Other numbers are exact by definition: e.g., 1 kilogram equals exactly 1000 grams. Exact numbers never limit significant figures in calculations.

"Rounding Off"

If we want to avoid accumulating errors due to our rounding off, we will save our rounding off for our final result.

If you are doing a series of calculations you can:

a. carry _all_ the extra digits through to the result, and then "round off," or:

b. carry at least one extra digit through each step, and then "round off."

(Some chemists indicate these "extra" uncertain digits by writing them below the line of significant digits: for example: 1.41_{7262}.)

1. Look at your answer: How many significant figures should it have? Which digit will be the first uncertain digit?

2. What is the digit just to the right of the first uncertain digit?

a. If this digit is less than 5, it is dropped and the first uncertain digit remains unchanged. (This is called "rounding down.")

b. If this digit is greater than 5, it is dropped, and the first uncertain digit is increased by 1. (This is called "rounding up.")

c. If this digit is equal to 5, the first uncertain digit remains unchanged if it is even, and is increased by 1 if it is odd. (That is, when the second uncertain digit is 5, we will "round up" half the time and "round down" half the time. Our rounding errors should average out.)

or c. When the _second_ uncertain digit is 5, "round" the first uncertain digit to an _even_ number.

Problem 1-6:

Round off each of the following numbers to three significant figures:

a. $.025047 \rightarrow .0250_{47} \rightarrow .0250$

The leading zero is _not_ significant.
4 is the second uncertain digit: round down.

b. $14.062 \rightarrow 14.0_{62} \rightarrow 14.1$

6 is the second uncertain digit: round down.

c. $1.4652 \rightarrow 1.46_{52} \rightarrow 1.46$

5 is the second uncertain digit:
6 is the first uncertain digit: "round even."

d. $38.752 \rightarrow 38.7_{52} \rightarrow 38.8$

    5 is the second uncertain digit:
    7 is the first uncertain digit: "round even."

Conversion Factors

We will use "conversions factors" in many different sorts of chemistry problems. We begin using "conversion factors" to convert units of measurements. We can use "dimensional analysis" (tracking our units) to <u>check</u> the setup for a problem; sometimes we can even use the units to guide us <u>through</u> <u>setting up</u> a problem.

As you set up any problem, check to see if the process "makes sense" (that is, the process agrees with your "intuition," and if the answer "makes sense").

Problem 1-7:

a. How many pounds does a 2.35 kg piece of cheese weigh?

$$? \text{ pounds} = 2.35 \text{ kg}$$

We need a "conversion factor" between pounds and kilograms.

We find:

$$.4536 \text{ kg} = 1 \text{ pound (lb.)}$$

$$2.205 \text{ lb.} = 1 \text{ kg}$$

b. If we know how many pounds 1 kilogram "weighs," we could multiply by 2.35 kg. That is, we need a factor of "pounds per kilogram" or lb/kg: 2.205 lb/kg.

$$? \text{ pounds} = 2.35 \text{ kg} \times \frac{2.205 \text{ lb}}{1.000 \text{ kg}}$$

$$= 5.18_{17} \text{ lb} \rightarrow 5.18 \text{ lb.}$$

c. <u>Check</u> that the units, when multiplied and divided, yield the right <u>units</u> for the answer:

$$\frac{\text{kg} \times \text{lb}}{\text{kg}} = \text{lb}$$

<u>That</u> is dimensional analysis.

The factor $\left(\dfrac{2.205 \text{ lb}}{1.000 \text{ kg}}\right)$ is called a "unit factor" because it doesn't change the mass of the cheese, just the units by which it is measured.

For this problem you had a choice of conversion factors:

What if you had used .4536 kg = 1 lb?

a. Determine how many pounds per kilogram:

$$\frac{1.000 \text{ lb}}{.4536 \text{ kg}}$$

b. Use this factor:

$$2.35 \text{ kg} \times \frac{1.000 \text{ lb}}{.4536 \text{ kg}} = 5.18_{07} \text{ lb} \rightarrow 5.18 \text{ lb}.$$

c. Check the units:

$$\frac{\text{kg} \times \text{lb}}{\text{kg}} = \text{lb}$$

Does the answer "make sense"? 1 kilogram is about 2 pounds: We'd expect 2.35 kg to be about 5 lbs. The answer does "make sense," if you include the units.

These steps seem (and are) very simple. If you take care, though, not to skip these simple steps, you won't be led astray in more complex problems.

To do more complex conversions, break the problem into steps.

Problem 1-8:

A car is traveling 55.0 miles per hour. How many feet per second does it travel?

Plan: miles → feet

         hours → minutes → seconds

         $\dfrac{\text{feet}}{\text{seconds}}$

Solution:

$$55.0 \text{ miles} \times \frac{5280 \text{ feet}}{1 \text{ mile}} = 290_{400} \text{ feet} = 2.90_4 \times 10^5 \text{ feet}$$

$$1 \text{ hour} \times \frac{60 \text{ min}}{1 \text{ hour}} \times \frac{60 \text{ s}}{1 \text{ min}} = 3600 \text{ s (exactly)}$$

$$\frac{2.90_4 \times 10^5 \text{ feet}}{3600 \text{ s}} = \frac{8.06_6 \times 10^1 \text{ feet}}{\text{s}} = \frac{80.7 \text{ feet}}{\text{s}}$$

Temperature Conversions

This approach doesn't rely on memorizing a formula. You can "think through" the process if you remember: 100 Celsius degrees are equivalent to 180 Fahrenheit degrees <u>and</u> 0.0° C equals 32.0° F (the freezing point of water).

Problem 1-9:

Convert 68.0° F to the corresponding temperature on the Celsius scale.

a. How far from 32.0° F is 68.0° F?   68.0° F − 32.0° F = 36.0° F,

That is, a temperature difference of 36.0° Fahrenheit degrees.

b. What temperature difference in °C does this correspond to?

$$36.0° \text{ Fahrenheit degrees} \times \frac{100° \text{ C}}{180° \text{ F}} = 20.0° \text{ Celsius degrees}$$

c. This temperature is 20.0 Celsius degrees above 0.00°C:

The temperature is 20.0°C.

Problem 1-10:

What is 32.5° C on the Fahrenheit scale?

a. 32.5° C is 32.5 Celsius degrees "above freezing" (0.0°C).

32.5 Celsius degrees = ? Fahrenheit degrees.

$$32.5 \text{ Celsius degrees} \times \frac{180 \text{ Fahr. degrees}}{100 \text{ Celsius degrees}} = 58.5 \text{ Fahr. degrees}$$

This temperature is 58.5 Fahrenheit degrees above 32.0° F:

The temperature is 90.5° F.

Kelvin degrees are the same size as Celsius degrees. Look at the scales in Figure 1.10.

$$K = 273.15 + °C.$$

CHAPTER ONE: SELF-TEST

The self-test questions provide a quick survey of your understanding of the concepts in the chapter. Answer these questions before you begin the problems at the end of the chapter in the text. These questions are an introduction to the material in the chapter. To master the material you must do the problems at the end of the chapter in the text.

The answers to the self-test questions follow these questions.

1-1.   <u>Without</u> <u>using</u> <u>your</u> <u>calculator</u>, complete the following calculations:

a. $\dfrac{(3 \times 10^2)(4 \times 10^3)}{(2 \times 10^4)} =$

b. $(3.0 \times 10^3) - (2 \times 10^2) =$

c. $(5 \times 10^1)(4 \times 10^{-2}) =$

d. $(4 \times 10^4) + (7 \times 10^4) =$

e. $\sqrt{2.5 \times 10^5} =$

f. $(3.0 \times 10^4)^2 =$

1-2.   <u>Using</u> <u>your</u> <u>calculator</u>, complete the following calculations:

a. $\dfrac{(6.245 \times 10^{-6})(7.472 \times 10^{+7})}{(3.162 \times 10^2)(5.378 \times 10^{-4})} =$

b. $(7.317 \times 10^{-5}) + (6.713 \times 10^{-4}) + (8.291 \times 10^{-5}) =$

c. $\sqrt{6.25 \times 10^{-4}}$

d. $(3.2 \times 10^{-3})^3$

e. $\left[\dfrac{5.867 \times 10^{-4}(3.710)}{6.200 \times 10^{-2}} \right]^{1/3}$

1-3.   a.   How may nanograms are there in a kilogram?

b.   How many micrometers are there in a centimeter?

c.   How many kilometers are there in a meter?

d.   How many milliliters are there in a microliter?

1-4.   How many significant figures are there in each of the following quantities?

a.   1355.4 g

b.   0.00572 m

c. 6.730×10^{-4} L

d.   5500

1-5.   Round each of the following numbers to <u>4</u> significant figures.

a.   0.61754

b.   8.32162

c.   0.0062712

d.   8.64457

e.   2.79151

1-6.   A car is travelling 55.0 miles per hour. What is its speed in feet per second?

1-7.   A page of paper measures 8.50 inches by 11.0 inches. What is the area of this paper in cm^2?

1-8.   Convert each of the following temperatures in °F to °C and to °K.

a.   68.0°F

b.   −20.0°F

c.   104.5°F

1-9.   The density of gold is 19.32 g/cm³

a.   What volume does 2.44 g of gold occupy?

b.   How much does 0.458 cm³ of gold weigh?

CHAPTER ONE: SELF-TEST ANSWERS

1-1. a. 60

     b. 2.8×10^3

     c. 2

     d. 1.1×10^5

     e. $500 = 5.0 \times 10^2$

     f. 9.0×10^8

1-3. a. 10^{12} ng/kg

     b. 10^4 μm/cm

     c. 10^{-3} km/m

     d. 10^{-3} mL/μL

1-5. a. 0.6175

     b. 8.322

     c. 0.006271

     d. 8.644

     e. 2.792

1-9. a. 0.126 cm^3

     b. 8.85 g

1-2. a. 2.744×10^3

     b. 8.274×10^{-4}

     c. 2.50×10^{-2}

     d. 3.3×10^{-8}

     e. $(3.511 \times 10^{-2})^{1/3} = .3274$

1-4. a. 5 sig. fig.

     b. 3 sig. fig.

     c. 4 sig. fig.

     d. It's hard to tell without a
context; if the author means
5.5×10^3 2 sig. fig.
5.500×10^3 4 sig. fig.

1-6. 80.7 ft/s

1-7. 93.5 in^2 = 603 cm^2

1-8. a. 20.0°C; 293.2 K (0°C = 273.15 K)

     b. -28.9°C; 244.3 K

     c. 40.3°C; 313.4 K

CHAPTER TWO: ATOMS, MOLECULES AND IONS

Chapter 2 begins with some historical background of the development of modern atomic theory.

Then you begin to write symbols for atoms and ions. Complete symbols include the symbol for the element, the number of protons in the nucleus, the number of nucleons in the nucleus, and the charge on the ion. Commonly chemists use a shorthand notation with only the symbol for the element and the charge on the ion.

Next, you begin to write names and formulae for compounds, even before you understand much about how compounds form.

Begin by memorizing the symbols and names for Elements 1 through 38

and   Cs, Ba, Ag, Pt, Au, Hg, Cd, Pb, Sn, I, Xe, U, Pu.

Don't memorize atomic numbers or masses

Your instructor may select others, but these are the most common.

Keep your periodic chart handy: become familiar with the locations of these elements; begin to collect information about families of elements: e.g., elements in column 7A form −1 ions.

Your objectives for Chapter 2 should include being able to:

Describe the results of:

> Thomson's cathode ray experiments
>
> Mullikan's oil drop experiment
>
> Rutherford, Geiger, and Marsden's alpha ray and metal foil experiments

Apply the law of conservation of mass

Apply the law of definite proportions

Demonstrate that data fit the law of multiple proportions

Use Avogadro's hypothesis to predict formulae for gaseous compounds

Explain how Dalton's atomic theory "explains" these three laws

Understand isotopes

Write a "complete" symbol for an atom of ion

Name ionic compounds

Name simple binary covalent (nonionic) compounds

Hints:

Begin to do problems as soon as possible.

Practice naming compounds and writing formulae for compounds you use in the laboratory.

The Laws of Conservation of Mass and the Law of Definite Proportions

If mass is conserved in a chemical process, the total mass of the reactants must equal the total mass of all of the products.

Problem 2-1:

67.26 grams of mercury (II) oxide (a red powder) is heated to produce 62.28 grams of mercury (a silver liquid) driving off oxygen gas.

a. How many grams of oxygen gas is driven off when this sample is heated?

Solution:

The total mass of the reactants (mercuric oxide) = the total mass of the products (mercury and oxygen)

or

$$67.26 \text{ g} = 62.28 \text{ g} + \text{mass oxygen}$$

$$\text{mass oxygen} = 67.26 \text{ g} - 62.28 \text{ g}$$

$$4.98 \text{ g} = \text{g oxygen produced}$$

b. How much mercury would react with 35.23 grams of oxygen to form mercuric oxide?

Solution:

The law of definite proportions says that elements combine in a fixed ratio (a definite proportion) to form a compound:

From part a: 62.28 g of mercury react with 4.98 g of oxygen

Method I:

We could set up a ratio:

$$\frac{62.28 \text{ g mercury}}{4.98 \text{ g oxygen}} = \frac{? \text{ g mercury}}{35.23 \text{ g oxygen}}$$

(The proportion of mercury reacting with oxygen)

$$X \text{ g mercury} = \frac{62.28 \text{ g mercury}}{4.98 \text{ g oxygen}} \times 35.23 \text{ g oxygen}$$

$$= 441 \text{ g mercury}$$

Method II:

If we know how much mercury reacts with 1.00 gram of oxygen, we could multiply to determine how much mercury would react with 35.23 g oxygen.

$$X \text{ g mercury} = \frac{62.28 \text{ g mercury}}{4.98 \text{ g oxygen}} \times 35.23 \text{ g oxygen}$$

$$= \frac{20.55 \text{ g mercury}}{1.00 \text{ g oxygen}} \times 35.23 \text{ g oxygen}$$

$$= 441 \text{ g mercury}$$

Notice that the mathematics is the same for both methods. For either method you have to develop your own problem solving style. We'll spend much more time doing problems with reactants and products in the next chapters.

Problem 2-2:

One liter of oxygen gas reacts with two liters of fluorine gas to form two liters of gaseous product (at the same temperature; pressure):

What is the formula for the product?

Solution:

Avogadro's hypothesis indicates that each liter of gas contains the same number of gas particles (molecules).

We need two liters of fluorine gas for each liter of oxygen gas: therefore, two fluorine molecules are required for each oxygen molecule. Each liter of oxygen gas produces two liters of product: therefore, one oxygen molecule reacts with two molecules of fluorine to produce two molecules of product.

Oxygen gas is diatomic; that is, oxygen gas has the formula O_2

Fluorine gas is also diatomic:

If we write an equation, we have:

$$O_2 + 2 \ F_2 \rightarrow 2 \text{ molecules of product}$$

Remember, atoms are only rearranged in chemical reactions, never created or destroyed. Therefore, there must be 2 oxygen atoms and 4 fluorine atoms on

23

the right hand side of our equation, and they must be divided between two molecules. Therefore:

$$O_2 + 2 \ F_2 \rightarrow 2 \ OF_2$$

Isotopes

Isotopes are atoms of the same element which differ in mass, that is, they differ only in the number of neutrons in their nuclei, and they have the same number of protons in their nuclei.

Problem 2-3:

Consider the following table

Atom	Protons	Neutrons	Electrons	Mass Number
I	6	6	6	12
II	6	7	6	13
III	7	7	7	14
IV	6	8	6	14
V	7	8	7	15

a. Which atoms are isotopes of carbon?

b. Which atoms are isotopes of nitrogen?

Solution:

Look at the periodic chart:

Carbon has an atomic number 6 and an average mass of about 12.

Nitrogen has an atomic number 7 and an average mass of about 14.

Isotopes of carbon will each have six protons: I, II, IV.

Isotopes of nitrogen will each have seven protons: III, V.

c. Write the symbols for each: $^{12}_{6}C$, $^{13}_{6}C$, $^{14}_{7}N$, $^{14}_{6}C$, $^{15}_{7}N$

The subscript is redundant, actually. The atomic symbol indicates the number of protons in the nucleus.

Ions

Ions form when an atom (or collection of atoms) gains or loses electrons (remember, the protons are held tightly in the nucleus and are never lost or gained in a chemical reaction).

To write complete symbols for atoms and ions remember:

mass number   Charge: usually omitted for atoms
(no. of neutrons +   (Protons – electrons)
 no. of protons)
                        atomic symbol
                                                   Z   ±
                                        or: X
atomic number   A
(no. of protons in
 the nucleus)

Problem 2-4:

An <u>atom</u> has 12 protons, and 13 neutrons in its nucleus:

a. How many electrons does the neutral atom have?   <u>12</u>

If the atom is neutral, there are as many electrons as protons.

b. Write the complete symbol for this atom:

If the atom has 12 protons, its atomic number is 12:

$$_{12}^{Z}X$$

Looking at the periodic chart; the element is magnesium:

$$_{12}^{Z}Mg$$

The atom has no net charge:

$$_{12}^{Z}Mg^{0}$$

There are 12 protons and 13 neutrons in the nucleus:

25 nucleons, so the mass number is 25

$$_{12}^{25}Mg^{0}$$

Problem 2-5:

An ion has 38 protons, 36 electrons, and 50 neutrons:

Write the complete symbol for this ion

25

Solution:

Looking at the periodic chart for the element for atomic number 38:

$$38 + ^{50}_{38} Sr$$

This ion was formed by a strontium atom losing 2 electons: there are 38 protons and 36 electrons.

Therefore the charge is +2.

$$^{88}_{38} Sr^{+2}$$

Naming Ionic Compounds

Name the positive ion first.

Name the negative ion last.

Cations

Remember, some metals form only one kind of monatomic ion:

Column 1A forms + 1 ions: for example, K^+: Potassium ion
Column 2A forms + 2 ions: for example, Ca^{2+}: Calcium ion

Transition elements often form more than one ion:

The charge of the ion is indicated by a Roman numeral in the na.ie: Fe^{+2}, iron (II) ion; Fe^{+3}, iron (III) ion

Anions

Column 7A forms −1 monatomic ions: Cl^-: Chloride ion
Column 6A forms −2 monatomic ions: S^{2-}: Sulfide ion

Monatomic anions are named with an −ide ending.

Problem 2−6:

Name the following ionic compounds

a. $CaCl_2$

b. K_2O

c. $FeCl_3$

d. Al_2O_3

Solution:

a. $CaCl_2$: Calcium chloride:

You don't need to indicate in the name how many chlorides there are for each calcium:

Calcium only forms +2 calcium ions
Chlorine only forms −1 chloride ions

Ionic compounds contain equal amounts of positive and negative charge: therefore, we need two (−1) anions for each (+2) cation.

b. K_2O: Potassium Oxide

c. $FeCl_3$: Iron is a transition element and can form more than one kind of ion. How do we know what the charge on the iron ion is?

There are three Cl^- ions (Why?); Therefore the iron ion must be Fe^{+3} (Why?)

Then, the name is iron (III) chloride

d. Al_2O_3: Aluminum Oxide

What is the charge on the aluminum ion?

There are three (−2) oxide ions, for a total negative charge of −6; There are two positively charged aluminum ions, for a total positive charge of +6. Therefore each aluminum ion is (+3).

Problem 2-7:

Write the formulas for the following ionic compounds

Chromium (III) oxide
Barium Nitrate
Calcium fluoride
Sodium sulfate

Cr_2O_3:

Each oxide ion is (−2), each chromium ion is (+3) in order to have as much positive charge as negative charge, we need 2 (+3) ions and 3 (−2) ions.

$Ba(NO_3)_2$:

The −ate ending indicates that nitrate is not a monatomic anion. We have to memorize some ions: Nitrate is NO_3^-. Barium is in column 2A and forms +2 ions.

CaF_2:

 Calcium is in column 2A and forms +2 ions.
 Fluorine is in column 7A and forms −1 ions.

Na_2SO_4:

 Sodium is in column 1A and forms +1 ions. We have to memorize the sulfate ion: $SO_4{}^{2-}$.

Covalent Compounds

We name binary (containing only two elements) covalent (non-ionic) compounds as though they are ionic.

 i) Name the "more metallic" element first: The "most metallic" elements are in the lower left hand corner of the periodic chart. The "least metallic" element is fluorine, in the upper right hand corner of the periodic chart.

Therefore, in a compound formed by oxygen and silicon, silicon is more metallic and will be named first.

 ii) Use the −ide ending for the "less metallic" element.

 iii) Because there are no ions which would determine the ratios of atoms present in a covalent compound, we use Greek prefixes (See Table 2.4) to indicate how many of each atom is present. We don't use the prefix mono-for the first element.

 Therefore,

 SiO_2 would be named silicon dioxide.

 CO would be carbon monoxide. (Carbon is "more metallic" than oxygen.)

 SF_6 sulfur hexafluoride

 ClF_3 Chlorine trifluoride (Chlorine is "more metallic" than fluorine).

 OF_2 Oxygen difluoride

 N_2O_4 Dinitrogen tetroxide

There are other molecules for which we use common names: For example, water: H_2O and ammonia: NH_3.

CHAPTER TWO: SELF-TEST

Try these short answer practice questions: remember, these are just to get you started: your exam questions and problems will be more like the problems in the text.

2-1.   Match each of the following experiments with a result.

   a.   Thomson's cathode ray experiments

   b.   Rutherford, Marsden and Geiger's α particle experiments

   c.   Millikan's oil drop experiments

   i)   The mass of an atom is concentrated in the nucleus

   ii)   The charge of an electron is 1.602×10^{-19} C

   iii)   The ratio of charge to mass of an electron is 1.758×10^{11} c/kg

   iv)   The electron is negatively charged.

2-2.   Match each of the following results with an appropriate law.

   a.   In one compound 8 g of oxygen combine with 6 g of carbon;
        in another 16 g of oxygen combine with 6 g of carbon.

   b.   In one sample of an oxygen-sodium compound 23 g of sodium combine
        with 8 g of oxygen; in another sample of the same compound 46 g of
        sodium combine with 16 g of oxygen.

   i)   The law of definite proportions

   ii)   The law of multiple proportions

   iii)   The law of conservation of mass

2-3.   The text indicates that the nucleus has a radius of about 10^{-13} cm, and
        the electrons are, on the average, 10^{-8} cm from the nucleus. If the
        nucleus were the size of a quarter (radius about 1 cm), on the same
        scale, how far would electrons be from the nucleus?

   i)   10 meters (a little more than 10 yards)

   ii)   100 meters (a little more than a football field)

   iii)   1000 meters (about 0.6 miles)

2-4. Write the symbol for each of the following atoms or ions:

    a. with atomic number 11, mass number 23 and 11 electrons

    b. with 35 protons, 45 neutrons, and 36 electrons

    c. with mass number 88, 50 neutrons, and 36 electrons

2-5. Name the following compounds using the systematic names.

    a. Li_2SO_4

    b. Al_2O_3

    c. N_2O_4

    d. HBr

    e. $FeCl_3$

    f. Cu_2O

    g. HgO

2-6. Write the formula for each of the following compounds:

    a. chromium(III)chloride

    b. aluminum sulfate

    c. calcium nitrate

    d. barium hydroxide

    e. sodium phosphate

2-7. One liter of gaseous element A combines with (reacts with) two liters of gaseous element B to form two liters of a gaseous product.

    a. Is gas A monatomic or diatomic? How do you know?

    b. Is gas B monatomic or diatomic? How do you know?

2-8. A student is repeating Millikan's oil drop experiment. She measures the charge on four different oil drops and finds:

Drop	Charge
#1	6.0×10^{-15} xtpls
#2	3.0×10^{-15} xtpls
#3	12.0×10^{-15} xtpls
#4	4.5×10^{-15} xtpls

From these experiments,

what is the charge on an electron in xtpls (a new unit for measuring charge).

CHAPTER TWO: SELF-TEST ANSWERS

2-1.   a.   iii (and iv)

       b.   i

       c.   ii

2-2.   a.   ii

       b.   i

2-3.   iii)   The electron's average distance from the nucleus is approximately 10^5 times the radius of the nucleus.

$$\frac{10^{-8}}{10^{-13}} = 10^5 \text{ cm} = 10^3 \text{ m}$$

2-4.   a. $^{23}_{11}Na$

       b. $^{80}_{35}Br^{-1}$

       c. $^{88}_{38}Sr^{+2}$

2-5.   a.   lithium sulfate

       b.   aluminum oxide

       c.   dinitrogen tetroxide

       d.   hydrogen bromide (or hydrobromic acid)

       e.   iron(III)chloride

       f.   copper(I)oxide

       g.   mercury(II)oxide

2-6. $CrCl_3$

      $Al_2(SO_4)_3$

      $Ca(NO_3)_2$

      $Ba(OH)_2$

      Na_3PO_4

2-7.   a.   Diatomic: one liter of A yields two liters of product

      b.   It could be either:

         If B is diatomic the reaction is:

$$A_2 + 2 \ B_2 \rightarrow 2 \ AB_2$$

         If B is monatomic the reaction is:

$$A_2 + 2 \ B \rightarrow 2 \ AB$$

2-8. 1.5×10^{-15} xptls.   Each charge is a multiple of 1.5×10^{-5} xptls.

CHAPTER THREE: STOICHIOMETRY

The word "stoichiometry" comes from the Greek words stoikheion (element) and metron (measure). Stoichiometry problems focus on the relationships among reactants and products, how much will react, what is left over, how much is produced. Central to these calculations is the concept that atoms are neither created, nor destroyed, in a chemical process, just rearranged. Another central concept is that of a mole, a sort of "chemist's dozen." A "baker's dozen" is thirteen, a "chemist's dozen" is 6.022 x 10^{23}, Avogadro's number.

If one atom of one element reacts with three atoms of another to form a molecule, then one dozen atoms of the first will react with three dozen atoms of the second, one "hundred" atoms of the first will react with three "hundred" atoms of the second, and one mole of atoms of the first will react with three moles of atoms of the second.

The mole that most chemists use is the "gram-mole." A gram-mole of carbon "weighs" 12.011 grams. Some engineers use pound-moles (a pound-mole of carbon "weighs" 12.011 pounds) and ton-moles (a ton-mole of carbon "weighs" 12.011 tons). (What would Avogadro's number be for a ton-mole?) Unless otherwise indicated, we will be using the gram-mole. (A gram is a unit of mass, not weight; however we will often use "weighs" to mean "has a mass of.")

Remember, there is a difference between a mole of atoms of a diatomic element (like oxygen, O_2) and a mole of molecules. A mole of oxygen atoms weighs 15.9994 grams; a mole of oxygen molecules weighs 31.9988 grams. A problem mentioning "a mole of oxygen gas" implies a mole of oxygen molecules. Otherwise, "a mole of oxygen" is ambiguous.

When we use moles of an ionic compound, we don't have moles of molecules. Instead, we have moles of "formula units." A mole of Na_2CO_3 contains two moles of sodium (Na^+) ions and one mole of carbonate (CO_3^{2-}) ions, one mole of Na_2CO_3 "formula units." As you will see in Chapter 4, the positive and negative ions are arranged in the solid in a regular way, but the positive ions are not attached to particular negative ions.

Types of problems you should be able to do:

- Calculate the average atomic mass of an element, given the abundance of its various isotopes.

- Convert from mass of an element or compound to moles of an element or compound and vice versa.

- Calculate the number of atoms or molecules of an element, or molecules of a compound, given the mass of a sample.

- Calculate the "percent composition" of a compound: the mass of each element present in 100 grams of compound.

- Given the "percent composition" of a compound, calculate the "empirical" (simplest) formula of the compound.

- Given the "empirical" formula of a compound, and its molecular weight, determine the molecular formula of the compound.

- Balance any (!) chemical equation, given the formulae of the reactants and the products.

- Calculate the moles of one reactant required to react with a given number of moles of the other.

- Calculate the mass of one reactant required to react with a given mass of the other.

- Calculate the amount of product produced when a given mass of reactant reacts in the presence of an excess of the other reactants.

- Calculate the amount of product produced when given masses of different reactants are allowed to react. (Which is the limiting reagent?)

As your study of chemistry goes on, more stoichiometry will be introduced: including calculations using volumes and concentrations of solutions, volumes of gases, or the amount of heat given off during a reaction.

Problem 3-1:

Chlorine occurs as two isotopes, ^{35}Cl and ^{37}Cl. Each ^{35}Cl "weighs" 34.969 amu; each ^{37}Cl "weighs" 36.966 amu. If chlorine is 75.53% ^{35}Cl and 24.47% ^{37}Cl, what is the average atomic mass of chlorine?

Solution:

First, let's estimate what the answer will be. Clearly, the average atomic mass must fall between 34.969 and 36.966. As more of the chlorine atoms have a mass of 34.969 amu, the average atomic weight must lie closer to 34.969 than to 36.966. Therefore, the average atomic mass must lie between 35 amu and 36 amu.

Next, let's map out a solution:

Out of every 100 atoms, 75.53 "weigh" 34.969 amu and 24.47 "weigh" 36.966 amu. (If you are uncomfortable about fractions of atoms, as well you might be, you could choose 10,000 atoms: 7553 with a mass of 34.969 amu and 2447 with a mass of 36.966 amu.) What is the average mass of a chlorine atom?

The total mass of 100 atoms is the mass of the ^{35}Cl atoms plus the mass of the ^{37}Cl atoms:

$$75.53 \text{ atoms} \times 34.969 \text{ amu/atom} + 24.74 \text{ atoms} \times 36.966 \text{ amu}$$

$$= 2641._{2086} \text{ amu} + 904.5_{5802} \text{ amu}$$

(Calculators persist in giving us more digits than are justified. Use care when rounding off.)

$$= 3545 \text{ amu}$$

The mass of one "average" atom is then:

$$\frac{3545 \text{ amu}}{100 \text{ atoms}} = 35.45 \text{ amu}.$$

(If you chose a sample of 10,000 atoms, the sample weighed 3.545×10^5 amu, and each "average" atom has a mass of 35.45 amu.)

Problem 3-2:

Carbon occurs naturally in two forms, graphite and diamond. How many grams of carbon does 37.25 moles of carbon weigh? How many atoms of carbon are there in this sample?

(Do you really need to know that carbon occurs as graphite and as diamond in order to solve this problem? Many problems will contain extraneous information: your challenge is to sort out the information that you need from all the information provided.)

Solution:

Given: moles of carbon.

To find: grams of carbon, and atoms of carbon.

Plan:

$$\text{moles of carbon} \xrightarrow[\text{atomic mass}]{} \text{grams of carbon}$$

$$\text{moles of carbon} \xrightarrow[\text{Avogadro's number}]{} \text{atoms of carbon}$$

Now we see that we need two other pieces of information:

The atomic mass of carbon: 12.011 amu or 12.011 grams/mole and

Avogadro's number: 6.022×10^{23}

To calculate the grams of carbon: we need to know how many grams per mole of carbon:

$$37.25 \text{ moles C} \times \frac{12.011 \text{ gram C}}{1 \text{ mole C}} = 447.4 \text{ grams C}$$

To calculate the number of atoms of carbon, we need to know how many atoms per mole of carbon:

$$37.25 \text{ moles C} \times \frac{6.022 \times 10^{23}}{1 \text{ mole C}} = 2.243 \times 10^{25} \text{ atoms C}$$

(Remember both significant figures and units for each answer.)

36

Problem 3-3:

How many atoms in 47.16 grams of sulfur?

Solution:

Given: the mass of sulfur

To be found: atoms of sulfur

Plan:

$$\text{mass S (grams)} \xrightarrow[\text{atomic mass S}]{} \text{moles S} \xrightarrow[\text{Avogadro's number}]{} \text{atoms S}$$

We can carry out the calculations for each step, or we can set up the entire problem, and then do the calculations.

Other information needed: atomic mass of sulfur: 32.06 amu or 32.06 grams/mole

Avogadro's number: 6.022×10^{23}

Checking units:

$$\text{grams of S} \times \frac{\text{moles of S}}{\text{grams of S}} \times \frac{\text{atoms of S}}{\text{moles of S}} = \text{atoms of S}$$

$$47.16 \text{ g S} \times \frac{1 \text{ mole S}}{32.06 \text{ grams S}} \times \frac{6.022 \times 10^{23} \text{ atoms S}}{1 \text{ mole S}}$$

$$= 1.471 \text{ moles S} \times 6.022 \times 10^{23} \text{ atoms S/mole}$$

$$= 8.858 \times 10^{23} \text{ atoms S.}$$

Do these answers make sense? 47.16 grams S is more than 1 mole, but not 2 moles. 1.471 moles seems a reasonable answer.

Suppose a student hurries through this problem and sets it up this way:

$$\frac{47.16 \times 32.06}{6.022 \times 10^{23}} = 2.510 \times 10^{-21} \text{ atoms}$$

Why should the student immediately see from the value that this answer isn't possible?

If the student had used units on each term, he would have grams^2/atom for the units of his answer.

Check each of your answers to see if it makes sense. Check the units on your answer. Both of these techniques will help you check for errors.

Problem 3-4:

What is the percent composition of sodium nitrate, $NaNO_3$?

Solution:

In one mole of sodium nitrate we have:

1 mole of sodium atoms

1 mole of nitrogen <u>atoms</u> and

3 moles of oxygen <u>atoms</u>

Given: the formula of the compound:

Plan: Calculate the mass of one mole of the compound:

$$\text{moles Na} \xrightarrow[\text{atomic mass Na}]{} \text{grams Na}$$

$$\text{moles N} \xrightarrow[\text{atomic mass N}]{} \text{grams N}$$

$$\text{moles O} \xrightarrow[\text{atomic mass O}]{} \text{grams O}$$

Total mass of compound (formula weight) = mass Na + mass N + mass O.

How much sodium is in one gram of compound?

Then we can calculate the percent of sodium (grams of the sodium in 100 grams of compound) by multiplying by 100:

$$\frac{\text{grams of Na}}{\text{mass of one mole of compound}} \times 100 \text{ grams compound} = \text{percent of Na}$$

Similarly for nitrogen and oxygen.

Ready?:

$$\frac{22.990 \text{ g Na}}{22.990 \text{ g Na} + 14.007 \text{ g N} + 3(15.999) \text{ g O}} = \text{g Na in one gram of compound}$$

$$= \frac{22.900 \text{ g Na}}{84.994 \text{ g compound}} = .27049 \text{ g Na/g compound}$$

.27049 g Na/g compound $\times$ 100 g compound = 27.049 percent Na

For nitrogen: there are 14.007 grams of nitrogen (1 mole of nitrogen atoms) in 84.994 g of compound (1 mole of compound):

$$\frac{14.007 \text{ g N}}{84.994 \text{ g compound}} \times 100 \text{ g compound} = 16.480 \text{ percent N}$$

For oxygen: there are 3(15.999) grams of oxygen (3 moles of oxygen atoms) in 84.994 g of compound (1 mole of compound):

$$\frac{3(15.999) \text{ g O}}{84.994 \text{ g compound}} \times 100 \text{ g compound} = 56.471 \text{ percent}$$

Check: The percent of Na plus the percent of N plus the percent of O should equal 100: 27.049 + 16.480 + 56.471 = ? (100.000; often we will find that rounding errors result in totals ranging from 99.9 to 100.1, depending on the number of significant figures we use.)

Problem 3-5:

A clear liquid organic compound is found to contain:

    52.2 percent carbon

    34.8 percent oxygen

    13.0 percent hydrogen

What is this compound's empirical (simplest) formula?

Solution:

The empirical formula of a compound indicates the ratios of moles of atoms of each element present in the compound. We can choose any size sample of our compound, and then determine the number of moles of atoms of each element present. If we are given percent composition, the most convenient sample is 100 g of compound: which contains 52.2 g of carbon, 34.8 grams of oxygen, and 13.0 grams of hydrogen.

$$\text{moles of C atoms} = \frac{52.2 \text{ g C}}{12.011 \text{ g C/mole C}} = 4.34 \text{ moles C}$$

or:

$$52.2 \text{ g C} \times \frac{1.000 \text{ mole C}}{12.011 \text{ g C}} = 4.34 \text{ moles C}$$

$$\text{moles of H atoms} = \frac{13.0 \text{ g H}}{1.008 \text{ g H/mole H}} = 12.89 \text{ moles H}$$

or:

$$13.0 \text{ g H} \times \frac{1.000 \text{ mole H}}{1.008 \text{ g H}} = 12.89 \text{ moles H}$$

$$\text{moles of O atoms} = \frac{34.8 \text{ g O}}{15.994 \text{ g O/mole O}} = 2.17_6 \text{ moles O}$$

or:

$$34.8 \text{ g O} \times \frac{1.000 \text{ mole O}}{15.994 \text{ g O}} = 2.17_6 \text{ moles O}$$

We are looking for the simplest whole number ratios of atoms in this compound. We can divide the number of moles of one element by the number of moles of any other. Usually we choose to divide by the smallest number of moles present:

$$\frac{\text{moles C}}{\text{moles O}} = \frac{4.35}{2.18} = 1.99_5 \text{ moles C/mole O} \approx 2$$

$$\frac{\text{moles H}}{\text{moles O}} = \frac{12.9}{2.18} = 5.91_7 \text{ moles H/mole O} \approx 6$$

Hints: Watch for ratios like 2.5 [= 5/2], 1.67 [= 5/3], 1.33 [= 4/3]: don't round these off.

The simplest formula then would be C_2H_6O. Remember, the simplest formula is not necessarily the molecular formula. The molecular formula could be C_2H_6O, $C_4H_{12}O_2$, $C_6H_{18}O_3$, $C_8H_{24}O_4$, etc. We could represent these possibilities as $(C_2H_6O)_x$, where x is an integer.

In order for us to determine the molecular formula, we need to know the molecular weight.

If the molecular weight of this compound is 46, what is the molecular formula?   (Hint: the molecular weight must be x times the weight of the simplest formula unit.)

The weight of the simplest formula unit is $2 \times 12 + 6 \times 1 + 1 \times 16 = 46$. Therefore, the empirical formula is, in this case, the molecular formula.

Problem 3-6:

If the simplest formula of an oxide of nitrogen is NO_2, and its molecular weight is approximately 90, what is the molecular formula of this nitrogen oxide?

Solution:

The weight of the simplest formula unit is $14 + 2 \times 16 = 46$. How many formula units are required to give a molecular weight of about 90?

90/46 is about 2:   Therefore the molecular formula contains 2 simplest formula units: the molecular formula is then $(NO_2)_2$ or N_2O_4.

Problem 3-7:

Balance the following equation:

$$C_3H_8O(\ell) + O_2(g) \rightarrow CO_2(g) + H_2O(\ell)$$

Initially, balancing equations is done by "inspection" (a sort of systematic trial and error), keeping these principles in mind:

- Don't change subscripts in the formulae: changing subscripts changes the compounds involved.

40

- Begin with the elements that occur in only one compound on each side of the reaction.

- Keep track of which coefficients you have fixed, and which remain to be determined.

- Check that each element is balanced.

Ready?

Solution:

$$\_\_ \; C_3H_8O(\ell) + \_\_ \; O_2(g) \rightarrow \_\_ \; CO_2(g) + \_\_ \; H_2O(\ell)$$

The blanks indicate that these subscripts have not yet been determined.

Begin with carbon and hydrogen:

$$\underline{1} \; C_3H_8O(\ell) + \_\_ \; O_2(g) \rightarrow \underline{3} \; CO_2(g) + \underline{4} \; H_2O(\ell)$$

Now, carbon and hydrogen are balanced, only oxygen remains unbalanced. On the right hand side of the equation we have 10 oxygen atoms. On the left hand side we have 1 oxygen atom from the C_3H_8O plus some from the O_2. How many O_2's do we need, to make a total of 10 oxygen atoms on the left hand side? hmmm: 9/2 O_2's. Our balanced equation could look like this:

$$\underline{1} \; C_3H_8O(\ell) + \frac{9}{2} \; O_2(g) \rightarrow \underline{3} \; CO_2(g) + \underline{4} \; H_2O(\ell)$$

However, we want whole number coefficients for our equation: we can multiply each coefficient by 2 to obtain:

$$\underline{2} \; C_3H_8O(\ell) + \underline{9} \; O_2(g) \rightarrow \underline{6} \; CO_2(g) + \underline{8} \; H_2O(\ell)$$

Now, check to ensure that each element is balanced:

	left hand side	right hand side
C atoms	6	6
H atoms	16	16
O atoms	2 + 18	12 + 8

Problem 3-8:

Calculate the moles of O_2 that would react with 5.84 moles of C_3H_8O according to the reaction in problem 3-7.

Solution:

Given: moles C_3H_8O

To find: moles O_2

Plan: moles C_3H_8O $\xrightarrow[\text{balanced equation}]{}$ moles O_2

We could: reason out the process.

or use ratios,

or look for a "conversion factor."

Each of these approaches is the same mathematically. Let's see how they compare. You will have to develop your own problem solving strategy.

Method I:

Reasoning out the process:

If we knew how many moles of O_2 were produced from one mole of C_3H_8O, we could calculate how many moles of O_2 were produced from 5.84 moles of C_3H_8O. The coefficients in the balanced equation tell us:

$$\frac{9 \text{ moles } O_2}{2 \text{ moles } C_3H_8O} \times 5.84 \text{ moles } C_3H_8O = 26.2_8 \text{ moles } O_2$$

Method II:

Using ratios:

$$\frac{9 \text{ moles } O_2}{2 \text{ moles } C_3H_8O} = \frac{x \text{ moles } O_2}{5.48 \text{ moles } C_3H_8O}$$

$$x = \frac{9 \text{ moles } O_2}{2 \text{ moles } C_3H_8O} \times 5.84 \text{ moles } C_3H_8O = 26.2_8 \text{ moles } O_2$$

Method III:

Using a "conversion factor": we are looking for a factor that has the units

$$\frac{\text{moles } O_2}{\text{moles } C_3H_8O}$$

Looking at the balanced equation we find:

$$\frac{9 \text{ moles } O_2}{2 \text{ moles } C_3H_8O}$$

Using our conversion factor we find:

$$\frac{9 \text{ moles } O_2}{2 \text{ moles } C_3H_8O} \times 5.84 \text{ moles } C_3H_8O = 26.2_8 \text{ moles } O_2$$

Each thought process leads us to the same mathematics. Whichever process we use we must remember to check that the units are correct. Use extra care if you use ratios to ensure that you haven't inverted one side of the ratio. A "ratio-thinking" student in a hurry might have set the ratio up this way:

$$\frac{9 \text{ moles } O_2}{x \text{ moles } O_2} = \frac{5.48 \text{ moles } C_3H_8O}{2 \text{ moles } C_3H_8O}$$

Although the units appear right, the right hand ratio is inverted, yielding an answer of 3.28 moles O_2. How does the student know that the answer is incorrect? We started with 5.48 moles of C_3H_8O. This is more than twice the two moles appearing in the balanced equation. Therefore, we should need more than twice the nine moles of O_2 appearing in the balanced equation.

Problem 3-9:

How many grams of C_3H_8O are required to react with 47.35 grams of O_2?

Solution:

Given: mass of O_2

To find: mass of C_3H_8O

Plan:

$$\text{Mass of } O_2 \xrightarrow[\text{mole wt } O_2]{} \text{moles of } O_2 \xrightarrow[\text{balanced equation}]{} \text{moles of } C_3H_8O$$

$$\xrightarrow[\text{mol wt } C_3H_8O]{} \text{mass of } C_3H_8O$$

$$47.35 \text{ g } O_2 \times \frac{1 \text{ mole } O_2}{32.00 \text{ g } O_2} \times \frac{2 \text{ moles } C_3H_8O}{9 \text{ moles } O_2} \times \frac{76.10 \text{ g } C_3H_8O}{1 \text{ mole } C_3H_8O}$$

$$= 25.10_2 \text{ grams } C_3H_8O$$

Problem 3-10:

How much CO_2 is produced when 13.74 grams of C_3H_8O are burned in excess oxygen?

Solution:

Given: mass of C_3H_8O

To find: mass of CO_2

Plan:

$$\text{mass of } C_3H_8O \xrightarrow[\text{mol wt } C_3H_8O]{} \text{moles of } C_3H_8O \xrightarrow[\text{bal'd equation}]{} \text{moles of } CO_2$$

$$\xrightarrow[\text{mol wt } CO_2]{} \text{mass of } CO_2$$

$$13.74 \text{ g } C_3H_8O \times \frac{1 \text{ mole } C_3H_8O}{76.10 \text{ g } C_3H_8O} \times \frac{6 \text{ moles } CO_2}{3 \text{ moles } C_3H_8O} \times \frac{44.01 \text{ g } CO_2}{1 \text{ mole } CO_2}$$

$$= 15.89 \text{ grams } CO_2$$

So far, all our stoichiometry problems fit one of the following pathways.

mass reactant $\xrightarrow[\text{mol wt}]{}$ moles reactant $\xrightarrow[\text{bal'd eqn}]{}$ moles product $\xrightarrow[\text{mol wt}]{}$ mass product

or:

mass reactant $\xrightarrow[\text{mol wt}]{}$ moles reactant $\xrightarrow[\text{bal'ed eqn}]{}$ moles of another reactant $\xrightarrow[\text{mol wt}]{}$ mass of another reactant

Later we will extend this map to include volumes of solutions and gases, heat given off, etc. Central to our map is the balanced equation.

The following problem contains a further complication.

Problem 3-11:

Aluminum reacts with oxygen gas to form aluminum oxide. How many grams of aluminum oxide will be formed when 22.45 grams of aluminum are burned with 21.72 grams of oxygen?

(Hint: Whenever a problem gives us amounts of <u>two</u> reactants, we need to determine the "limiting reagent," that is, the reactant which we will "run out of" first. This limits the amount of product that can be formed.)

We might be tempted to choose oxygen as the limiting reagent, as we have fewer grams of oxygen. Don't yield to this temptation!

We can take two approaches:

- How much aluminum oxide can be produced from the given amount of aluminum; how much aluminum oxide can be produced from the given amount of oxygen? The smaller amount is the amount of aluminum oxide we can produce. Or,

- How much oxygen is required to react with the given aluminum? Do we have enough? If we do, aluminum is the limiting reagent, and we can determine how much aluminum oxide is formed. If we don't have enough oxygen to react completely with the aluminum, oxygen is the limiting reagent and we can determine how much aluminum oxide is formed.

In each case, we need to carry out two stoichiometry problems. Choose the approach you like best.

Begin by balancing the equation:

$$4\ Al(s) + 3\ O_2(g) \rightarrow 2\ Al_2O_3(s)$$

Approach I:

i) How much aluminum oxide can be produced from 22.45 grams of aluminum?

mass Al $\xrightarrow[\text{at wt}]{}$ moles Al $\xrightarrow[\text{bal'd eqn}]{}$ moles Al_2O_3 $\xrightarrow[\text{mol wt}]{}$ grams Al_2O_3

$$22.45 \text{ g Al} \times \frac{1 \text{ mole Al}}{26.98 \text{ g Al}} \times \frac{2 \text{ moles Al}_2\text{O}_3}{4 \text{ moles Al}} = .4160 \text{ moles Al}_2\text{O}_3$$

ii)   How much aluminum oxide can be produced from 21.72 grams of oxygen?

$$\text{mass O}_2 \xrightarrow[\text{at wt}]{} \text{moles O}_2 \xrightarrow[\text{bal'd eqn}]{} \text{moles Al}_2\text{O}_3 \xrightarrow[\text{mol wt}]{} \text{grams Al}_2\text{O}_3$$

(Remember, we need moles of O_2 molecules, not mole of O atoms.)

$$21.72 \text{ g O}_2 \times \frac{1 \text{ mole O}_2}{32.00 \text{ g O}_2} \times \frac{2 \text{ moles Al}_2\text{O}_3}{3 \text{ moles O}_2} = .4525 \text{ moles Al}_2\text{O}_3$$

We have enough aluminum to make .4160 moles of aluminum oxide, and enough oxygen to make .4525 moles of aluminum oxide. Therefore, aluminum is the limiting reagent, and we can calculate the mass of aluminum oxide formed:

.4160 moles $Al_2O_3 \times$ 101.96 gram/mole = 42.42 grams aluminum oxide.

Approach II:

How much oxygen is required to react with 22.45 grams of aluminum?

$$\text{mass Al} \rightarrow \text{moles Al} \rightarrow \text{moles O}_2 \rightarrow \text{grams O}_2$$

$$22.45 \text{ g Al} \times \frac{1 \text{ mole Al}}{26.98 \text{ g Al}} \times \frac{3 \text{ moles O}_2}{4 \text{ moles Al}} \times \frac{32.00 \text{ g O}_2}{1 \text{ mole O}_2} = 19.97 \text{ g O}_2$$

Therefore, we will have oxygen left over, and aluminum is the limiting reagent. To calculate the amount of aluminum oxide produced:

$$\text{mass Al} \xrightarrow[\text{at wt}]{} \text{moles Al} \xrightarrow[\text{bal'd eqn}]{} \text{moles Al}_2\text{O}_3 \xrightarrow[\text{mol wt}]{} \text{grams Al}_2\text{O}_3$$

$$22.45 \text{ g Al} \times \frac{1 \text{ mole Al}}{26.98 \text{ g Al}} \times \frac{2 \text{ moles Al}_2\text{O}_3}{4 \text{ moles Al}} = .4160 \text{ moles Al}_2\text{O}_3$$

.4160 moles $Al_2O_3 \times$ 101.96 grams/mole = 42.42 grams aluminum oxide.

Sometimes the problem will also ask "how much of which reagent is left over?"

Once we know the limiting reagent, we can determine how much of the other reagent is used, and by subtraction, determine how much is left over. In this case: aluminum is the limiting reagent, 19.97 grams of oxygen are consumed, and:

21.72 grams O_2 initially present – 19.97 grams O_2 consumed

= 1.75 grams O_2 left

Limiting reagent problems can appear intimidating: but you can break them down into much simpler problems. Map out your plan before you begin.

3-1.   Naturally occurring boron consists of two isotopes, $^{10}_5B$ and $^{11}_5B$.   If 19.60% of the boron atoms are $^{10}_5B$, each weighing 10.013 amu and 80.40% of the boron atoms are $^{11}_5B$, each weighing 11.009 amu, what is the average mass of a boron atom?

3-2.   An atom weighs 2.414 times as much as a $^{12}_6C$ atom.   What is the mass of this atom in amu?   In grams?

3-3.   a.   How many moles of nitrogen <u>atoms</u> are there in 1.874 g of nitrogen gas?

   b.   How many moles of nitrogen <u>molecules</u> are there in 4.976 g of nitrogen gas?

   c.   How many <u>atoms</u> of nitrogen are there in 0.389 g of nitrogen gas?

   d.   How many <u>molecules</u> of nitrogen are there in 0.216 g of nitrogen gas?

   e.   How much does 0.847 moles of nitrogen gas weigh?

3-4.   What is the molecular weight of $C_2H_4O_2$?

3-5.   10.00 g of a compound contains 4.74 g carbon, 4.21 g oxygen and 1.05 g of hydrogen.   What is the simplest formula for the compound?

3-6.   What is the percent composition of $C_4H_{10}O$?

3-7.   Balance the following equations:

   a.   ___ H_2 + ___ Cl_2 → ___ HCl

   b.   ___ C_3H_8O + ___ O_2 → ___ CO_2 + ___ H_2O

   c.   ___ Al + ___ O_2 → ___ Al_2O_3

3-8.   Given the balanced equation:

$$2\ C_2H_6(g) + 7\ O_2(g) \rightarrow 4\ CO_2(g) + 6\ H_2O(g)$$

   a.   How many <u>molecules</u> of H_2O are produced when 10 molecules of C_2H_6 react with <u>excess</u> O_2?

   b.   How many <u>moles</u> of CO_2 are produced when 0.164 moles of C_2H_6 are burned in <u>excess</u> O_2?

c. How many moles of H_2O are produced when 0.427 moles of O_2 react with excess C_2H_6?

d. How many moles of H_2O are produced when 0.150 moles of C_2H_6 react with excess O_2?

e. How many moles of H_2O are produced when a mixture of 0.427 moles of O_2 and 0.150 moles of C_2H_6 is burned?

3-9. Ammonia (NH_3) reacts with oxygen gas (O_2) to give NO and water. How many grams of O_2 are required to react completely with 18.2 grams of NH_3?

3-10. A solution containing 6.130 g of NaCl is mixed with a solution containing 12.26 g of $AgNO_3$. AgCl precipitates.

a. How many grams of AgCl precipitates?

b. How many grams of excess reactant remain in solution?

(Hint: i. How many moles of AgCl can be produced from 6.130 g of NaCl?

   ii. How many moles of AgCl can be produced from 12.26 g of $AgNO_3$?)

CHAPTER TWO: SELF-TEST ANSWERS

3-1.   10.81 amu

3-2.   28.97 amu (1 amu = 1.6605×10^{-24} g)

4.81×10^{-23} g

3-3.   a.   0.1338 moles N atoms

b.   0.1776 moles N_2 molecules

c. 1.67×10^{22} atoms N

d. 4.64×10^{21} molecules N_2

e.   23.7 g

3-4.   60.05 g/mole

3-5. $C_3H_8O_2$

3-6.   64.82% C, 13.60% H, 21.58% O

3-7.   a. $H_2 + Cl_2 \rightarrow 2\ HCl$

b. $2\ C_3H_8O + 9\ O_2 \rightarrow 6\ CO_2 + 8\ H_2O$

c. $4\ Al + 3\ O_2 \rightarrow 2\ Al_2O_3$

3-8.   a.   30 molecules of H_2O

b.   0.328 moles CO_2

c.   0.366 moles H_2O

d.   0.450 moles H_2O

e. O_2 is the limiting reagent: 0.366 moles H_2O are produced.

3-9. $4\ NH_3 + 5\ O_2 \rightarrow 4\ NO + 6\ H_2O$

1.07 moles NH_3

42.8 g O_2

3-10.   i) 0.1049 moles AgCl

       ii) 0.07217 moles AgCl

       a. $AgNO_3$ is the limiting reagent; 10.34 g AgCl

       b. NaCl is the excess reagent; 0.07217 moles of NaCl react;
           4.218 grams NaCl react; 1.912 grams NaCl remain.

CHAPTER FOUR: SOLUTIONS

Chapter 4 introduces solutions, principally aqueous (water) solutions. You'll meet many new terms: Don't memorize their definitions, but do understand and apply the concepts.

A Recap of Concepts:

A solution is a homogeneous mixture (that is, a mixture with uniform composition throughout). We often think of solutions as having a solid (or a liquid) dissolved in a liquid. However, our definition would also include a gas "dissolved" in a gas, a gas dissolved in a liquid, a liquid dissolved in a solid (e.g., mercury dissolved in gold).

The solvent is usually the component present in the highest concentration. (We usually consider the solvent to be the component of the solution that has the same physical state as the solution; that is, when we make a liquid solution of sugar and water, we think of water as the solvent because water and the solution are both liquids.)

The solute is dissolved in the solvent.

Cations are positively (puss-i-tively? Chemists are noted for bad puns) charged ions; anions are negatively charged.

Ionic compounds are made up of arrays of cations and anions.

Polar molecules have no net charge, but one end of the molecule is more polar than another. Polar molecules act as dipoles (having two separated centers of charge. We often represent them as $(\delta+ \dots \delta-)$ where δ stands for a fraction of the charge of an electron.)

Non-polar molecules do not act as dipoles.

Some compounds dissolve in water, producing ions. Because these solutions can carry electricity, these compounds are called "electrolytes." A compound which dissociates partially is called a weak electrolyte.

Compounds which dissociate completely to give hydrogen ions (H^+) in water are strong acids. There are, of course, weak acids. Strong bases dissociate completely to give hydroxide ions (OH^-).

We can describe the composition of a solution by using the concentrations of each solute. The concentration is given by the amount of solute (grams, moles, etc.) for a given amount (liters, grams, etc.) of solvent or solution.

You will often use molarity (moles of solute per liter of solution) in the lab.

Chapter 4 introduces a new way of writing equations for reactions taking place in solution, the net ionic equation.

The net ionic equation focusses our attention on the ions and molecules undergoing a chemical change. Ions which do not undergo a chemical change are spectator ions.

In Chapter 3 we might have written:

$$NaCl(aq) + AgNO_3(aq) \rightarrow AgCl(s) \downarrow + NaNO_3(aq)$$

$NaCl$, $AgNO_3$ and $NaNO_3$ are strong electrolytes. Therefore, we might write this reaction as:

$$Na^+(aq) + Cl^-(aq) + Ag^+(aq) + NO_3(aq) \rightarrow AgCl \downarrow + Na^+(aq) + NO_3^-(aq)$$

Na^+ ions and NO_3^- ions are spectators; the net ionic equation is then:

$$Cl^-(aq) + Ag^+(aq) \rightarrow AgCl(s) \downarrow$$

Silver chloride precipitates from ("falls out of") this solution.

You should be able to write net ionic equations for

    a)   acid-base reactions

    b)   precipitation reactions

and c)   redox reactions

"Redox" is "short-hand" for reduction-and-oxidation. When a substance loses electrons it is oxidized; when a a substance gains electrons it is reduced. When sodium metal reacts with chlorine gas to form sodium chloride, sodium loses electrons to form sodium (+) ions, while chlorine gains electrons to form chloride (-) ions. Sodium is oxidized; chlorine is reduced.

In a redox electron, all the electrons lost by one reactant must be gained by another. We can use this rule to help us balance redox reactions: electrons lost = electrons gained. We'll need a bookkeeping method for keeping track of the electrons.

After this somewhat lengthy recap, let's do some problems:

We can add to our problem-solving "map" from Chapter 3:

$$grams \xrightarrow[MW]{} moles \xrightarrow[molarity]{} volume\ of\ solution$$

Remember, molarity = $\dfrac{moles\ solute}{liter\ solution}$ or moles of solute per liter of solution.

$$M = \frac{n}{V}$$

Problem 4-1:

a. How many moles of NaCl are in 74.5 mL of a 0.250 M (molar; moles/liter) solution?

Solution:

Convert 74.5 mL to liters → .0745 L

$$\left[74.5 \text{ mL} \times \frac{1.00 \text{ L}}{1000 \text{ mL}} \right] = .0745 \text{ L}$$

$$\text{moles NaCl} = 0.0745 \text{ L} \times \frac{0.250 \text{ moles}}{\text{L}} \text{ NaCl}$$

$$= .0186 \text{ moles} = 1.86 \times 10^{-3} \text{ moles}$$

Remember significant figures.

b. What volume of this solution would you measure out to obtain 0.147 moles of NaCl?

Solution:

Method I.

If you are comfortable rearranging $M = \frac{n}{V}$

then $V = \frac{n}{M}$ $\left[V \times m = \frac{nV}{V} \qquad V \times \frac{m}{m} = \frac{n}{m} \right]$

$$V = \frac{0.147 \text{ moles NaCl}}{0.250 \text{ moles NaCl/liter}} = 0.588 \text{ liters}$$

Method II.

If you aren't yet comfortable with that, consider

$$V \text{ (in liters)} = n \text{ (in moles)} \times ?$$

If we knew how many liters contained 1.00... mole, we could calculate how many would contain n moles.

From the molarity: 1.000 liters contains 0.250 moles

$$\frac{1.000 \text{ L}}{0.250 \text{ moles}} = \frac{4.00 \text{ liters}}{\text{mole}}$$

∴ 4.00 L contains 1.000 moles of NaCl:

$$V = 0.147 \text{ moles NaCl} \times \frac{1.000 \text{ L}}{0.250 \text{ moles}} = 0.588 \text{ liters}$$

Once you've thought this through, you'll see that method I is just short-cut, with the same mathematics.

Some problems will use mL for volume, mg for mass, and mmol (millimole). Remember, if a solution is 0.357 M (moles/liter), it is also 0.357 mmol/-mL. Convince yourself how many mmol of NaCl are in 32.6 mL of a 0.357 M solution?

$$32.6 \text{ mL} \times \frac{0.357 \text{ mmol NaCl}}{\text{mL}} = 11.6 \text{ mmol NaCl}$$

Now, convert the volume to liters and use the concentration of NaCl in moles/liter.

Solution Stoichiometry

We use the balanced equation as we did in Chapter 3 to calculate amounts of reactants and products and we can "expand" our "map" to include solutions.

$$g \text{ A} \xrightarrow{MW} \begin{array}{c} \text{moles} \\ \text{A} \end{array} \xrightarrow{\begin{array}{c} \text{balanced} \\ \text{equation} \end{array}} \begin{array}{c} \text{moles} \\ \text{B} \end{array} \xrightarrow{MW} g \text{ B}$$

Vol. A $\nearrow$ M.   M. $\searrow$ Vol. B

Problem 4-2:

What volume of 0.150 M KOH is required to precipitate <u>all</u> of the Cu^{2+} ion in 250.0 mL of 0.183 M $Cu(NO_3)_2$?

First, write (or find) the balanced equation:

$$Cu^{2+} + 2 \text{ OH}^- \rightarrow CuO \downarrow + H_2O$$

Map out a plan:

moles Cu^{2+} → moles OH^- → volume KOH (L)
or mmoles   or mmols   or (mL)

$$\text{moles } Cu^{2+} = .2500 \text{ L} \times \frac{0.183 \text{ moles}}{\text{L}}$$

$$= 4.575 \times 10^{-2} \text{ moles } Cu^{2+}$$

The balanced equation indicates each mole of Cu^{2+} requires 2 moles of OH^-.

$$\text{moles } OH^- = 2 \times \text{moles } Cu^{2+}$$

$$\text{moles } OH^- = \frac{2 \text{ moles } OH^-}{1 \text{ mole } Cu^{2+}} \times (4.57_5 \times 10^{-2} \text{ moles } Cu^{2+}) = 9.15 \times 10^{-2} \text{ moles } OH^-$$

$$\text{volume OH}^- = \frac{\text{moles OH}^-}{M_{OH^-}} = \text{moles OH}^- \times \frac{1.000 \text{ L OH}^-}{.150 \text{ moles OH}^-}$$

$$= \frac{9.15 \times 10^{-2} \text{ moles OH}^-}{0.150 \text{ M}} = 0.610 \text{ L}$$

Problem 4-3:

A 25.00 mL aliquot (measured portion) of a 0.1342 M $AgNO_3$ solution is diluted with water to a final volume of 500.0 mL in a volumetric flask. What is the concentration (molarity) of the dilute $AgNO_3$?

In this dilution problem, remember that the number of moles of $AgNO_3$ in the 500.0 mL solution equals the number of moles of $AgNO_3$ in the original 25.00 mL sample.

$$\text{moles} = \text{molarity} \times \text{volume (liters)}$$

$$[\text{or mmol} = \text{molarity} \times \text{volume (mL)}]$$

$$n = M \times V$$

$$n \text{ AgNO}_3 = 0.1342 \ M \times 25.00 \text{ mL}$$

$$= 3.355 \text{ mmol}$$

Then, the final concentration is:

$$M = \frac{3.355 \text{ mmol}}{500.0 \text{ mL}} = \frac{6.71 \times 10^{-3} \text{ mmol}}{\text{mL}}$$

$$= 6.71 \times 10^{-3} \ M$$

Acid-Base Reactions

Acid-base reactions can be thought of as "proton transfer" reaction. For example:

$$\underset{\text{(ammonia)}}{NH_3} + H_2O \rightleftharpoons \underset{\text{(ammonium ion)}}{NH_4^+} + OH^-$$

$$\underset{\text{(acetic acid)}}{CH_3COOH} + H_2O \rightleftharpoons \underset{\text{(acetate ion)}}{CH_3COO^-} + H_3O^+$$

$$NH_3 + CH_3COOH \rightleftharpoons NH_4^+ + CH_3COO^-$$

Notice: Not every hydrogen atom in a molecule is "acidic," and not every compound containing hydrogen is an acid.

You'll study more about acids in bases in Chapter 14.

For now, we'll use balanced equations to do stoichiometry problems.

Problem 4-4:

How much (what volume of) 0.273 M hydrochloric acid (HCl) is required to react completely with 50.00 mL of 0.167 M barium hydroxide ($Ba(OH)_2$)?

a. Balance the equation:

$$2 \ HCl(aq) + Ba \ (OH)_2(aq) \rightarrow BaCl_2(aq) + H_2O$$

b. Plan:

$$\text{Volume BaOH}_2 \xrightarrow{M_{Ba(OH)_2}} \text{moles Ba(OH)}_2 \xrightarrow[\text{equation}]{\text{balanced}} \text{moles HCl} \xrightarrow{M_{HCl}} \text{Volume HCl}$$

The acid loses a proton and the base accepts a proton.

c. Moles $Ba(OH)_2$ = 0.0500 liters $\times \dfrac{.167 \text{ moles}}{\text{liter}}$

$$= 8.35 \times 10^{-3} \text{ moles}$$

From the balanced equation:

$$\text{moles HCl} = 2 \times \text{moles Ba(OH)}_2$$

$$\text{moles HCl} = 2 \times (8.35 \times 10^{-3}) = 1.67 \times 10^{-2} \text{ moles}$$

To find the volume of HCl:

$$M = \frac{n}{V} = \frac{\text{moles}}{\text{liters}}$$

Rearranging: $\quad V_{HCl} = \dfrac{n_{HCl}}{M_{HCl}} = \dfrac{1.67 \times 10^{-2} \text{ moles}}{0.273 \text{ moles/liter}}$

$$= 6.12 \times 10^{-2} \text{ liters}$$

$$\left[\text{or} \qquad V = \text{moles HCl} \times \frac{\text{liters}}{\text{mole}} \right.$$

$$\left. = 1.67 \times 10^{-2} \text{ moles} \times \frac{1.00 \text{ L}}{0.273 \text{ moles}} \right]$$

Problem 4-5:

If you mix 35.5 mL of 0.148 M $AgNO_3$ with 10.0 mL of 0.320 M $BaCl_2$, silver chloride precipitates. Which reagent is limiting? How much (grams) AgCl precipitates?

a. Begin with the balanced equation:

$$2 \ AgNO_3(aq) + BaCl_2(aq) \rightarrow 2 \ AgCl(s) + Ba(NO_3)_2(aq)$$

b. Plan:

$$Vol.\ AgNO_3 \rightarrow moles\ AgNO_3 \rightarrow moles\ AgCl$$

which is smaller?

$$Vol.\ BaCl_2 \rightarrow moles\ BaCl_2 \rightarrow moles\ AgCl$$

c. Moles $AgNO_3$ = 0.354 L × 0.148 moles/liter = 5.24×10^{-3} moles $AgNO_3$. This $AgNO_3$ would produce 5.24×10^{-3} moles AgCl <u>if</u> we had enough barium chloride to precipitate all the silver ion.

moles $BaCl_2$ = 0.0100 L × 0.320 moles/liter = 3.20×10^{-3} moles $BaCl_2$

Don't be tempted to think that $BaCl_2$ is limiting yet! Use the balanced equation to calculate how much AgCl would precipitate if we had enough silver nitrate to precipitate <u>all</u> the chloride ion.

$$3.20 \times 10^{-3}\ moles\ of\ BaCl_2 \times \frac{2\ moles\ AgCl}{1\ moles\ BaCl_2} = 6.40 \times 10^{-3}\ moles\ of\ AgCl$$

We have enough silver nitrate to produce 5.24×10^{-3} moles of AgCl and enough barium chloride to produce 6.40×10^{-3} moles of AgCl.

<u>Therefore</u>: $BaCl_2$ is in excess, $AgNO_3$ is the limiting reagent, and 5.24×10^{-3} moles of AgCl precipitate.

To finish:

$$\rightarrow moles\ AgCl \rightarrow grams\ AgCl$$

$$5.24 \times 10^{-3}\ moles\ AgCl \times \frac{143.32\ g\ AgCl}{mole} = 0.751\ grams\ AgCl$$

Hint: Use your stoichiometry "map" to break each problem into simpler steps; Then follow your map.

This problem is much like the limiting reagent problems in Chapter 3.

Balancing Redox Reactions

Many redox reactions can be balanced "by inspection" or "by trial and error." However, we can use a systematic approach. There are two related methods.

Method of Oxidation States

Use the rules in Table 4.4 to assign oxidation states. (You might wonder how to know which of two elements has "greater attraction for the electrons in a bond." We'll discuss this in Chapter 8. For now, I'll just say that electron-attracting-ability <u>increases</u> to the <u>right</u> in the periodic chart, from column IA through column VIIA, and <u>increases</u> up a column in the periodic chart. Therefore, fluorine has the <u>greatest</u> ability to attract electrons in a bond.)

Assigning oxidation states is a "bookkeeping method" for electrons. We'll look at oxidation states again in Chapter 8.

Try to balance this skeleton equation:

$$\_\_ \ MnO_2(s) + \_\_ \ HCl(aq) \rightarrow MnCl_2(aq) + \_\_ \ Cl_2(g) + \_\_ \ H_2O(\ell)$$

i) First assign oxidation states. Remember, oxygen is −2 in compounds (except for peroxides) and hydrogen is +1 in compounds (except for metal hydrides).

MnO_2 Mn must be + 4, as the <u>sum</u> of the oxidation states
$+4 + (-2)2 = 0$ for one Mn and two O's <u>must</u> equal zero.

HCl
$+1 + (-1) = 0$

$MnCl_2$
$+2 + (-1)2 = 0$

Cl_2 Elements have a zero oxidation state.
0

H_2O
$2(+1) + (-2) = 0$

ii) Connect the atoms which change oxidation state

$$\overset{+4}{MnO_2} + \underset{-1}{\_\_ \ HCl} \rightarrow \underset{-1}{\_\_ \ \overset{+2}{MnCl_2}} + \underset{0}{\_\_ \ Cl_2} + H_2O$$

Notice that some of the chlorines change oxidation state, and others don't:

$Mn^{+4} \rightarrow Mn^{+2}$ A decrease of 2

$Cl^{-1} \rightarrow Cl^0$ An increase of 1

iii) Balance the <u>changes</u> in oxidation states:

We need two chlorine atoms changing oxidation state for each manganese atom: therefore 1 Cl_2 for every MnO_2 that reacts:

$$\underline{1} \ MnO_2 + \_\_ \ HCl \rightarrow \underline{1} \ MnCl_2 + \underline{1} \ Cl_2 + \_\_ \ H_2O$$

iv) Now we can finish balancing the equation:

The oxygen atoms must balance, the chlorine atoms must balance, and the hydrogen atoms must balance.

To balance the chlorine atoms: choose the coefficient of the HCl:

$$\underline{1} \ MnO_2 + \underline{4} \ HCl \rightarrow \underline{1} \ MnCl_2 + \underline{1} \ Cl_2 + \_\_ \ H_2O$$

Then balance the oxygen atoms: choose the coefficient of water:

$$\underline{1}\ MnO_2 + \underline{4}\ HCl \rightarrow \underline{1}\ MnCl_2 + \underline{1}\ Cl_2 + \underline{2}\ H_2O$$

Now check the hydrogen atoms: they are balanced.

<u>The half-reaction method</u> can be used for any redox reaction, although we usually use it for reactions in solution. It is another book-keeping method.

Let's apply this method to the reaction we've just balanced, using the net ionic equation.

$$\underline{}\ MnO_2 + \underline{}\ H^+ + \underline{}\ Cl^- \rightarrow \underline{}\ Mn^{2+} + \underline{}\ Cl_2 + \underline{}\ H_2O$$

i) Identify the half-reactions:

$$MnO_2 \rightarrow Mn^{2+}$$

$$Cl^- \rightarrow Cl_2$$

ii) Balance each half-reaction

a. First, balance for all elements but hydrogen and oxygen

$$MnO_2 \rightarrow Mn^{2+} \qquad\qquad 2\ Cl^- \rightarrow Cl_2$$

b. Then, balance for oxygen with H_2O

$$MnO_2 \rightarrow Mn^{2+} + 2\ H_2O \qquad\qquad 2\ Cl^- \rightarrow Cl_2$$

c. Next, balance for hydrogen with H^+

$$4\ H^+ + MnO_2 \rightarrow Mn^{2+} + 2\ H_2O \qquad\qquad 2\ Cl^- \rightarrow Cl_2$$

d. Then, balance for charge with electrons:

(Add electrons to the more positive side of each half-reaction.)

$$4\ H^+ + MnO_2 \rightarrow Mn^{2+} + 2\ H_2O \qquad\qquad 2\ Cl^- \rightarrow Cl_2$$

$$\text{net} + 4 \qquad \text{net} + 2 \qquad\qquad \text{net} -2 \qquad \text{net } 0$$

Therefore:

$$4\ H^+ + MnO_2 + 2e^- \rightarrow Mn^{2+} + 2\ H_2O \qquad\qquad 2\ Cl^- \rightarrow Cl_2 + 2e^-$$

MnO_2 gains electrons and is reduced.

HCl (Cl^-) loses electrons and is oxidized.

iii) Combine the half-reactions, so that the number of electrons produced (by Cl^-) equals the number of electrons consumed (by MnO_2).

In this case: we just add the two half-reactions:

$$4\ H^+ + MnO_2 + 2e^- \rightarrow Mn^{2+} + 2\ H_2O$$

$$2\ Cl^- \rightarrow Cl_2 + 2e^-$$

$$\overline{}$$

$$4\ H^+ + MnO_2 + 2\ Cl^- + 2e^- \rightarrow Mn^{2+} + 2\ H_2O + Cl_2 + 2e^-$$

iv) If we add two Cl^- ions as "spectators" to each side, we have:

$$4\ H^+ + MnO_2 + 4\ Cl^- \rightarrow Mn^{2+} + 2\ H_2O + Cl_2 + 2\ Cl^-$$

or $\qquad\qquad 4\ HCl + MnO_2 \rightarrow MnCl_2 + 2\ H_2O + Cl_2$

Problem 4-6:

Balance:

$$\underline{}\ ClO_3^- + \underline{}\ SO_3^{2-} \rightarrow \underline{}\ Cl^- + SO_4^{2-}$$

Choose your method:

The oxidation state method

ClO_3^-
$+5\ (-2)3$ | The sum of the oxidation states equals the charge on the ion.

Cl^-
-1

SO_3^{2-}
$+4 +(-2)3 = -2$

SO_4^{2-}
$+6 + (2)4 = -2$

$$\overset{\text{decrease 6}}{\underset{}{\overline{}}}$$

$$\overset{+5}{\underline{}\ ClO_3^-} + \underset{+4}{\underline{}\ SO_3^{2-}} \rightarrow \overset{-1}{\underline{}\ Cl^-} + \underset{+6}{\underline{}\ SO_4^{2-}}$$

$$\underset{\text{increase 2}}{\overline{}}$$

We need 3 SO_3^{2-} ions to be oxidized for every 1 ClO_3^- to be reduced.

$$\underline{1}\ ClO_3^- + 3\ SO_3^{2-} \rightarrow \underline{1}\ Cl^- + \underline{3}\ SO_4^{2-}$$

Check to see that all atoms are balanced and that charge is balanced.

Balancing by half-reactions:

$$\underline{?} \; ClO_3^- \rightarrow \underline{?} \; Cl^- \qquad\qquad \underline{?} \; SO_3^{2-} \rightarrow \underline{?} \; SO_4^{2-}$$

$$ClO_3^- \rightarrow Cl^- + \underline{?} \; H_2O \qquad\qquad \underline{?} \; H_2O + SO_3^{2-} \rightarrow SO_4^{2-}$$

$$\underline{?} \; H^+ + ClO_3^- \rightarrow Cl^- + 3 \; H_2O \qquad H_2O + SO_3^{2-} \rightarrow SO_4^{2-} + \underline{?} \; H^+$$

$$\underline{?} \; e^- + 6 \; H^+ + ClO_3^- \rightarrow Cl^- + 3 \; H_2O$$

$$H_2O + SO_3^{2-} \rightarrow SO_4^{2-} + 2 \; H^+ + \underline{?} \; e^-$$

$$6 \; e^- + 6 \; H^+ + ClO_3^- \rightarrow Cl^- + 3 \; H_2O \qquad H_2O + SO_3^{2-} \rightarrow SO_4^{2-} + 2 \; H^+ + 2 \; e^-$$

Combine the half-reactions

$$3(H_2O + SO_3^{2-} \rightarrow SO_4^{2-} + 2 \; H^+ + 2 \; e^-)$$

$$6 \; e^- + 6 \; H^+ + ClO_3^- \rightarrow Cl^- + 3 \; H_2O$$

$$6 \; e^- + 6 \; H^+ + ClO_3^- + 3 \; H_2O + 3 \; SO_3^{2-} \rightarrow 3 \; SO_4^{2-} + 6 \; H^+ + 6 \; e^- + Cl^- + 3 \; H_2O$$

Cancelling from both sides yields:

$$ClO_3^- + 3 \; SO_3^{2-} \rightarrow 3 \; SO_4^{2-} + Cl^-$$

These two method of balancing redox reactions will let you balance very complex redox reactions.

Once you've balanced these equations you can, of course, do stoichiometry calculations with them.

CHAPTER FOUR: SELF-TEST

4-1.   Classify each of the following aqueous solutions as

      i)   a strong electrolyte (a good conductor of electricity)

      ii)   a weak electrolyte

or iii)   a nonelectrolyte

   a. NH_3 d. H_2O g. $Ba(NO_3)_2$

   b.   HCl   e.   NaCl

   c. $C_6H_{12}O_6$ (glucose)   f.   HF

4-2.   Calculate the molarity of the following solutions:

   a.   0.740 moles $Ba(OH)_2$ in 400.0 mL of solution

   b.   32.76 g NaCl in 0.750 L of solution

   c.   25.00 mL of 0.4521 M HCl diluted with water to 325.00 mL

4-3.   What is the concentration of each ion in the following solutions:

   a.   0.164 M $BaCl_2$

   b.   0.287 M HNO_3

   c.   1.61 M $(NH_4)_2SO_4$

4-4.   Use the solubility rules to decide if each of the following compounds is soluble or insoluble in water:

   a. $NaNO_3$ d. $PbSO_4$ g.   KOH

   b.   AgI   e. $BaCO_3$

   c. $(NH_4)_2SO_4$ f. Na_2S

4-5.   How many grams of $BaSO_4$ precipitate when 250.0 mL of 0.100 M $Ba(NO_3)_2$ is mixed with excess Na_2SO_4?

4-6.   How could you separate a mixture of Ag^+ ions, Ba^{2+} ions, and Cu^{2+} ions?

4-7.   How many mL of 0.1247 M NaOH are required to neutralize 25.00 mL of 0.1862 M HCl?

4-8.   How many mL of 0.1369 M NaOH are required to completely neutralize 32.00 mL of 0.1181 M H_2SO_4?

4-9. Assign oxidation numbers to each atom in the following compounds:

a. NH_4Cl d. $BaSO_4$

b. $CrCl_3$ e. $KMnO_4$

c. Al_2O_3 f. $K_2Cr_2O_7$

4-10. Balance the following oxidation-reduction reactions.

a. __ Cu + __ Ag^+ → __ Cu^{2+} + __ Ag

b. __ Cu + __ NO_3^- + __ H^+ → __ Cu^{2+} + __ NO + __ H_2O

c. __ I^- + __ $Cr_2O_7^{2-}$ + __ H^+ → __ Cr^{3+} + __ I_2 + __ H_2O

4-11. In each of the following reactions, identify the species oxidized and the species reduced.

$$5 H_2S + 2 MnO_4^- + 6 H^+ → 5 S + 2 Mn^{+2} + 8 H_2O$$

$$2 Na + 2 H_2O → 2 NaOH + H_2 \uparrow$$

$$2 C_2H_6 + 7 O_2 → 4 CO_2 + 6 H_2O$$

CHAPTER FOUR: SELF-TEST ANSWERS

4-1.   a.   weak   d.   nonelectrolyte   g.   strong

       b.   strong   e.   strong

       c.   nonelectrolyte   f.   weak

4-2.   a.   1.85 M

       b.   0.747 M

       c.   0.03478 M

4-3.   a. $[Ba^{2+}]$ = 0.164 M $[Cl^-]$ = 0.328 M

       b. $[H^+]$ = 0.287 M $[NO_3^-]$ - 0.287 M

       c. $[NH_4^+]$ = 3.22 M $[SO_4^{2-}]$ = 1.61 M

4-4.   a.   soluble   d.   insoluble   g.   soluble

       b.   insoluble   e.   insoluble

       c.   soluble   f.   soluble

4-5. 2.50×10^{-2} moles $BaSO_4$; 5.83 g $BaSO_4$

4-6.   Add a solution of NaCl to precipitate AgCl

      Then, add a solution of Na_2SO_4 to precipitate $BaSO_4$, leaving Cu^{2+} in solution.

4-7.   37.33 mL NaOH

4-8.   55.21 mL NaOH

4-9.   a.   N -3; Cl -1; H +1   d.   Ba +2; S +6; O -2

       b.   Cr +3; Cl -1   e.   K +1; Mn +7; O -2

       c.   Al +3; O -2   f.   K +1; Cr +6; O -2

4-10. a. $Cu + 2 Ag^+ \rightarrow Cu^{2+} + 2 Ag$

       b. $3 Cu + 2 NO_3^- + 8 H^+ \rightarrow 3 Cu^{2+} + 2 NO + 4 H_2O$

       c. $6 I^- + Cr_2O_7^{2-} + 14 H^+ \rightarrow 2 Cr^{3+} + 3 I_2 + 7 H_2O$

4-11. a. H_2S is oxidized; MnO_4^- is reduced

       b.   Na is oxidized; H_2O is reduced

       c. C_2H_6 is oxidized; O_2 is reduced

CHAPTER FIVE: GASES

Pressure

The definition of pressure is "force per unit area."   That is,

$$\text{Pressure} = \frac{\text{Force}}{\text{Area}}$$

In the System Internationale, the unit of force is the Newton (N).   How big is a Newton?   A 145 lb. (65.8 kg) human standing on the earth exerts a force of about 645 N;   1.00 ounces of cheese (on the earth) exerts a force of 0.278 N.   Therefore, a Newton is <u>not</u> a large force.

In the System Internationale, the unit of pressure is the Pascal (Pa), a pressure of 1 N/m². One Pascal, one Newton of force spread over an area of one square meter, is not a large pressure.   At sea level, the atmospheric pressure is about 10^5 Pa.

Measuring Pressure

Why is the height of the column of mercury in a barometer proportional to the atmospheric pressure?   Look at Figure 5.1 in the text: the pressure exerted by the atmosphere equals the pressure exerted by the column of mercury.

What is the pressure exerted by the column of mercury?   Pressure = Force/ Area.   Let's let the cross-sectional area of the tube be A.   How can we calculate the force?   The force (F) exerted by the column of mercury is the mass (m) of mercury multiplied by the acceleration of gravity (g).   We can look up the acceleration of gravity.   How can we calculate the mass of the mercury?   The mass (m) of mercury equals the volume (v) of the mercury times the density (d) of mercury.   We can look up the density of the mercury.   How can we calculate the volume of the mercury column?   The volume of a cylinder is the area (A) × the height (h).

Then the volume of the column of mercury is: $V = h \cdot A$;

The mass of the column of mercury is: $m = d \cdot V = d \cdot h \cdot A$

The force exerted by the mercury is: $F = m \cdot g = (d \cdot h \cdot A) \cdot g$

The pressure exerted by the mercury is: $P = \dfrac{F}{A} = (d \cdot g)h$

So we've shown that the atmospheric pressure is <u>proportional</u> to the height of the mercury column it supports.   A pressure which supports 1 mm mercury Hg is defined as 1 torr, which equals 133.3 Pa.

We can convert from torr to Pascals, but most chemists continue to use torr as a unit of pressure.

Chemists also use the atmosphere (atm) as a unit of pressure.

$$1 \text{ atm} = 760 \text{ torr}$$

We can use a U-shaped tube to measure pressure differences:

atmosphere

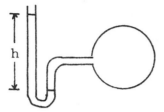

The pressure of the gas in the container is h mm greater than the pressure exerted by the atmosphere. If the atmospheric pressure is measured with a barometer and found to be 745 mm Hg (torr), then the pressure of the gas in the box is (745 + h) torr.

Boyle's Law

"The volume of a gas is inversely proportional to its pressure." Or, more simply, the pressure of a gas times its volume is a constant (at a given temperature).

We can write: $P \cdot V = k$ (constant)

$$\text{or} \quad V = k\left(\frac{1}{P}\right)$$

Real gases do not follow Boyle's Law exactly. However Boyle's Law is a good approximation for gases near 1 atmosphere pressure.

Problem 5-1:

A 2.50 liter tank contains oxygen gas at 5.75 atm pressure. What volume will this oxygen occupy at a pressure of 0.836 atm (atmospheric pressure in Denver, Colorado)?

Boyle's Law: $P_{initial} \cdot V_{initial} = k$

If the temperature doesn't change, $P_{final} \cdot V_{final} = k$, also.

Then $P_i \cdot V_i = P_f \cdot V_f$

We know the initial pressure, the initial volume, and the final pressure.

$$(5.75 \text{ atm}) \cdot (2.50 \text{ L}) = (0.836 \text{ atm}) \cdot V_f$$

$$V_f = (2.50 \text{ L}) \left(\frac{5.75 \text{ atm}}{0.836 \text{ atm}}\right) = 17.1_9 \text{ L} = 17.2 \text{ L}$$

Note: Dimensional analysis won't help catch all of our mistakes in this sort of problem. Does the answer make sense? As the pressure decreases, the volume should increase, which agrees with our calculation.

Note: In order to use Boyle's Law both pressures must be in the same units, and both volumes must be in the same units.

Charles' Law

The volume of a gas is directly proportional to the absolute temperature (in K) at a given pressure:

$$V = b \cdot T$$

using the symbol T for "absolute temperature."

This equation can be rearranged to give:

$$\frac{V}{T} = b,$$

where b is constant at a given pressure. For the same sample of gas, we can calculate the final volume from the initial volume and temperature, and the final temperature.

$$\frac{V_{initial}}{T_{initial}} = \frac{V_{final}}{T_{final}}$$

Problem 5-2:

A sample of gas occupies 3.62 L at 15°C and 0.85 atm pressure. What volume would this gas occupy a 45°C and 0.85 atm pressure?

$$\frac{V_{initial}}{T_{initial}} = \frac{V_{final}}{T_{final}}$$

Solution:

Remember that these temperatures must be in K!

$$T_i = 15°C + 273.15 = 288 \text{ K}$$

$$T_f = 45°C + 273.15 = 318 \text{ K}$$

$$\frac{3.62 \text{ L}}{288 \text{ K}} = \frac{V_f}{318 \text{ K}}$$

$$V_f = \frac{3.62 \text{ L}}{288 \text{ K}} \cdot 318 \text{ K} = 3.99_7 \text{ L} = 4.00 \text{ L}$$

Does this answer make sense? As the temperature increases, the volume increases.

Avogadro's Law

At a given temperature and pressure, the volume of a gas is proportional to the number of moles (n) of the gas.

$$V = n \cdot a \quad (a \text{ is a constant})$$

$$\text{or} \quad \frac{V}{n} = a$$

Problem 5-3:

If 0.750 moles of nitrogen occupies 16.8 L at a given temperature and pressure, what volume would 1.24 moles of nitrogen occupy (at the same temperature and pressure)?

$$\frac{V_1}{n_1} = \frac{V_2}{n_2}$$

$$\frac{16.8 \text{ L}}{0.750 \text{ moles}} = \frac{V_2}{1.24 \text{ moles}}$$

$$V_2 = \frac{16.8 \text{ L}}{0.750 \text{ moles}} \times 1.24 \text{ moles} = 27.7_7 \text{ L} = 27.8 \text{ L}$$

[What volume would 1.24 moles of _hydrogen_ occupy at this temperature and pressure?

Equal volumes of gases at the same temperature and pressure contain equal numbers of moles of gas. Therefore, 1.24 moles of hydrogen would occupy the same volume as 1.24 moles of nitrogen (at the same temperature and pressure).]

The Ideal Gas Law

We can describe a sample of gas by its volume (V), its temperature (T), its pressure (P), and the number of moles of gas (n) present.

We've seen three gas laws:

Boyle's Law	$V = \dfrac{k}{P}$	(T and n constant)
Charles' Law	$V = bT$	(P and n constant)
Avogadro's Law	$V = a \cdot n$	(P and T constant)

Each focusses on how two properties behave when other properties are held constant.

There is a way to combine these laws:

$$V = \left(\frac{n \cdot T}{P} \right) \cdot \text{constant}$$

This constant is found to be the same for all gases:

we'll use the R to represent this constant:

$$V = R \cdot \frac{nT}{P}$$

This can be rearranged to give: $PV = nRT$

or $R = \dfrac{PV}{nT}$

(Check that you can "derive" each of the other laws from this one.)

The size and units for R depend on the units of pressure and volume you use. If pressure is measured in atmospheres and volume in liters:

$$R = 0.08206 \ \frac{L \ atm}{K \ mol}$$

We can use the Ideal Gas Law to find the number moles of a gas from its pressure, volume, and temperature.

We can add to our stoichiometry map from Chapter 3 (with additions from Chapter 4):

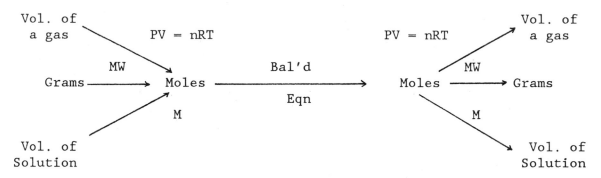

We can also use the Ideal Gas Law to solve problems when more than two properties vary at a time.

Problem 5-4:

A sample of hydrogen gas occupies 4.26 L at a pressure of 1.93 atm and a temperature of 38°C. What volume will this sample occupy if the pressure is reduced to 1.27 atm and the temperature is increased to 55°C?

Solution:

Method I.

We can break the problem into two steps.

   i. First change the pressure, and find V_2: Use Boyle's Law

68

ii. Then change the temperature, and find V_3: use Charles' Law

$$V_1 \to V_2 \to V_3$$

i. The pressure changes:

$$P_1V_1 = P_2V_2$$

$$1.93 \text{ atm} \cdot 4.26 \text{ L} = 1.27 \text{ mat} \cdot V_2$$

$$V_2 = \frac{1.93 \text{ atm}}{1.27 \text{ atm}} \cdot 4.26 \text{ L}$$

$$= 6.47_3 \text{ L}$$

ii. Then temperature changes

$$\frac{V_2}{T_2} = \frac{V_3}{T_3}$$

(We need temperatures in Kelvins.)

$$\frac{6.47_3 \text{ L}}{(273 + 38)K} = \frac{V_3}{(273 + 55)K}$$

$$V_3 = \frac{6.47_3 \text{ L}}{311 \text{ K}} (328 \text{ K}) = 6.82_6 \text{ L} \to 6.83 \text{ L}$$

Method II.

We can do the problem in one step:

Using the Ideal Gas Equation:

$$PV = nRT$$

i. Collect all the properties that change on one side of the equation and collect all the properties that remain constant on the other side. In this problem, pressure, volume and temperature change, while the moles of gas (n) remain unchanged. R is always constant.

$$\frac{PV}{T} = nR$$

Then:

$$\frac{P_{initial} \cdot V_{initial}}{T_{initial}} = \frac{P_{final} \cdot V_{final}}{T_{final}}$$

$$\frac{1.93 \text{ atm} \times 4.26 \text{ L}}{311 \text{ K}} = \frac{1.27 \text{ atm} \times V_{final}}{328 \text{ K}}$$

$$V_{final} = \frac{1.93 \text{ atm}}{1.27 \text{ atm}} \cdot \frac{328 \text{ K}}{311 \text{ K}} \cdot 4.26 \text{ L}$$

$$= 6.82_7 \text{ L} = 6.83 \text{ L}$$

When we do the problem in two steps, we can easily check each step to see if the answer to that step makes sense. How do we check our "one step" solution.

Let's look at the equation:

$$\frac{P_i \cdot V_i}{T_i} = \frac{P_f \cdot V_f}{T_f}$$

Solve for V_f:

$$V_f = \frac{P_i V_i}{T_i} \cdot \frac{T_f}{P_f}$$

Rearrange this equation:

$$V_f = \left(\frac{P_i}{P_f}\right) \cdot \left(\frac{T_f}{T_i}\right) V_i$$

We could think of (P_i/P_f) as a "pressure correction factor,"

and (T_f/T_i) as a "temperature correction factor" multiplied times the initial volume.

Then we can check to see whether each "correction factors" "makes sense" when we calculate V_{final}:

$$\frac{P_i}{P_f} = \frac{1.93 \text{ atm}}{1.27 \text{ atm}} :$$

The pressure decreases, and this factor increases the volume.

$$\frac{T_f}{T_i} = \frac{328 \text{ K}}{311 \text{ K}} :$$

The temperature increases and this factor increases the volume.

Both "correction factors" do make sense.

The molar volume of a gas is the volume one mole of the gas occupies. If n moles of gas occupy a volume V, the molar volume is V/n. From the Ideal Gas Law:

$$PV = nRT \qquad V = \frac{nRT}{P} \qquad \frac{V}{n} = \frac{RT}{P}$$

The molar volume of a gas clearly depends on the temperature and pressure. Chemists use 0°C (273 K) and 1 atm pressure as "standard temperature and pressure (STP)."

Problem 5-5:

What volume does 1.00 mole of gas occupy at STP?

$$V = \frac{nRT}{P} = \frac{1.00 \text{ mole} \left[\dfrac{.08206 \text{ L atm}}{\text{K mol}}\right]}{1.00 \text{ atm}} \times 273 \text{ K}$$

$$= 22.4 \text{ L}$$

Check units: $\dfrac{\text{mole} \times \dfrac{\text{L atm}}{\text{K mol}} \times \text{K}}{\text{atm}}$

If the conditions of a problem specify STP, you can use the molar volume as 22.4 L. However, you can always use PV = nRT.

Molecular Weight of a Gas

Problem 5-6:

The density of a gas is found to be 6.18 g/L at 25°C and 2.13 atm pressure. What is the molecular weight of the gas?

Solution:

If we knew how many moles of gas were in one liter at this pressure and temperature, we could calculate the molecular weight.

$$MW = \frac{\text{grams}}{\text{moles}}$$

Use PV = nRT to find the number of moles in 1.00 L

$$n = \frac{PV}{RT} = \frac{2.13 \text{ atm} \ (1.00 \text{ L})}{0.8206 \dfrac{\text{L atm}}{\text{K mol}} \ (298 \text{ K})}$$

$$= 0.0871 \text{ moles (check units)}$$

6.18 g of this gas occupies 1.00 L and is .0871 moles.

Then the molecular weight is:

$$MW = \frac{6.18 \text{ g}}{.0871 \text{ moles}} = \frac{70.9_5 \text{ g}}{\text{moles}}$$

$$\rightarrow 71.0 \text{ g/mole}$$

Dalton's Law of Partial Pressures can be expressed as

$$P_{total} = P_1 + P_2 + P_3 + \ldots$$

Where P_1 is the pressure exerted by gas 1 in a mixture, P_2 is the pressure exerted by gas 2, etc. and P_{total} is the total pressure exerted by the mixture of gases.

Therefore, when we write $PV = nRT$, n is the total number of moles of all gases in a mixture.

The Kinetic Molecular Theory of Gases is a fairly simple model of matter at a "microscopic" level (atoms and molecules) which predicts the behavior of matter (a gas) at a "macroscopic" level. KMT treats gas molecules as "tiny" particles in constant motion, bouncing against the walls of the container, and colliding with each other. This model predicts the Ideal Gas Law (including Boyle's Law, Charles' Law and Avogadro's Law), if we are willing to assume that the absolute temperature is a measure of the average "microscopic" kinetic energy of the gas molecules.

A more detailed view of the gas molecules (see Figure 5.14 and 5.15 in the text) shows the molecules do not all travel at the same speed; some travel much faster than the average, some much slower.

$$\text{If} \qquad (KE)_{AVG} = \frac{3}{2} RT$$

Then the average kinetic energy of gas molecules depends only on the temperature. Oxygen molecules (O_2) and hydrogen molecules (H_2) will all have the same average kinetic energy at the same temperature. Which molecules will travel at a higher speed?

Lighter molecules travel faster than heavier molecules at the same temperature. How do we know?

The kinetic energy for a particle is given by:

$$K.E. = \frac{1}{2} mu^2 \qquad \text{where u is the speed of the particle}$$

For a collection of identical particles, each with the same mass

$$\overline{K.E.} = \frac{1}{2} \overline{mu^2}, \qquad \text{Where the bar indicates average.}$$

$\overline{u^2}$ is the average square speed, not the square of the average speed. This distinction is a bit subtle.

An aside:

Consider 3 particles traveling at 2.0 m/s, 3.0 m/s, and 4.0 m/s.

i) What is the average speed of these particles?

$$\frac{2 + 3 + 4}{3} \frac{m}{s} = \frac{3.0\ m}{s}$$

ii) What is the square of the average speed?

$$\frac{9.0\ m^2}{s^2}$$

iii) What is the average square speed?

The square speeds are:

$$\frac{4.0\ m^2}{s^2}, \quad \frac{9.0\ m^2}{s^2}, \quad and \quad \frac{16.0\ m^2}{s^2}$$

The average square speed is then:

$$\frac{(4.0 + 9.0 + 16.0)m^2}{3\ s^2} = \frac{29.0\ m^2}{3\ s^2} = \frac{9.7\ m^2}{s^2}$$

iv) The "root mean square speed" of these particles is the square root of the average square speed:

$$\sqrt{\frac{9.7\ m^2}{s^2}} = \frac{3.1\ m}{s}$$

Therefore, for a collection of particles traveling at different speed, the average speed, 3.0 m/s, is not equal to the "root mean square" speed, 3.1 m /s.

Now,

the average speed of molecules in a gas can be calculated from the Kinetic Molecular Theory.

The average speed is given by:

$$\bar{u} = \sqrt{\frac{8\ RT}{\pi N_a m}}$$

If we compare this expression to the expression for the "root mean square speed" of the gas molecules:

$$u_{rms} = \sqrt{\bar{u^2}} = \sqrt{\frac{3\ RT}{N_a m}}$$

We can see that both expressions indicate that molecules travel faster at higher temperatures and that, at the same temperature, heavier molecules travel more slowly than lighter ones. Both u_{urms} and u are directly proportional to the square root of T and inversely proportional to the square root of the mass of the molecule.

If we are comparing two molecules, such as H_2 and O_2 at the same temperature

$$\frac{u_{rms\ H_2}}{u_{rms\ O_2}} = \sqrt{\frac{m_{O_2}}{m_{H_2}}}$$

m_{O_2}/m_{H_2} is the ratio of the mass of one O_2 molecules to that of one H_2 molecule. This ratio is the same as the ratio of the mass of a mole of oxygen molecules to the mass of a mole of hydrogen molecules

$$\text{and} \quad \frac{u_{H_2}}{u_{O_2}} = \sqrt{\frac{M_{O_2}}{M_{H_2}}}$$

$$\sqrt{\frac{M_{O_2}}{M_{H_2}}} = \sqrt{\frac{32.00}{2.00}} = 4.00$$

If the rates of effusion and diffusion are proportional to the average speeds of the molecules, we could calculate the relative rates of effusion and diffusion of hydrogen and oxygen:

$$\frac{\text{Relative Rate}_{H_2}}{\text{Relative Rate}_{O_2}} = \sqrt{\frac{m_{O_2}}{m_{H_2}}} = \sqrt{\frac{M_{O_2}}{M_{H_2}}} = 4.00$$

5-1.   Given the following conversion factors:

   1 atm = 760.0 torr   1 atm = 101,325 Pa

   Convert the following pressures:

   a.   570.5 torr to atm

   b.   1.24 atm to Pascals

   c. 2.37×10^5 Pa to torr

5-2.   A sample of hydrogen gas occupies 10.69 L at 2.48 atm at a given temperature.

   a.   What volume will this gas occupy at 1.68 atm at the same temperature?

   b.   At what pressure will this gas occupy 8.37 L at the same temperature?

5-3.   A sample of methane gas, CH_4, occupies 5.82 L at 52°C at a given pressure.

   a.   What volume will this gas occupy at 104°C at the same pressure?

   b.   At what temperature will this gas occupy 5.32 L at the same pressure?

5-4.   At a certain temperature and pressure, 2.40 moles of N_2 occupy 15.2 L.

   a.   What volume will 1.96 moles N_2 occupy at this temperature and pressure?

   b.   What volume will 1.60 moles of hydrogen gas, H_2, occupy at this temperature and pressure?

   c.   How many moles of N_2 will occupy 8.64 L at this temperature and pressure?

5-5.   Find the volume of CO_2 at 351°C and 0.852 atm when 48.5 g of $CaCO_3$ is heated to completely drive off carbon dioxide:

$$CaCO_3(s) \xrightarrow{\Delta} CO_2(g) + CaO(s)$$

5-6.   A 2.500 L sample of gas at 0.951 atm and 25°C weighs 2.924 grams.

   a.   What is the molecular weight of the gas?

   b.   If the empirical formula of this gas is CH_3, what is its molecular formula?

5-7. In a sample of air, 78.1% of the molecules are N_2 molecules; 20.9% of the molecules are O_2 molecules; 1.0% of the molecules are Ar molecules. If the total pressure of this sample of air is 645 mm Hg, what is:

a. the partial pressure of N_2

b. the partial pressure of O_2

c. the partial pressure of Ar

5-8. Compare two 28.0 L tanks of gas. Tank A contains 2.50 moles of H_2 at 273 K; Tank B contains 2.50 moles of O_2 at 273 K.

a. Which tank has the greater pressure?

b. Which molecules have the greater average kinetic energy?

c. Which molecules have the greater average speed?

5-9. A tank of gas contains a mixture of helium and oxygen.

a. Which gas escapes faster from a pin-hole leak in the tank?

b. How much faster does the faster gas escape?

5-10. For an ideal gas PV = nRT. We can derive this equation of state assuming that gas molecules are "point particles" with no attractive forces between them. Real gas molecules do occupy a real volume and do have attractive forces between them.

a. How do the attractive forces between molecules affect the pressure of a real gas?

b. How does the real volume of gas molecules affect the pressure of a real gas?

CHAPTER FIVE: SELF-TEST ANSWERS

5-1. a. 0.7506 atm

b. 1.26×10^5 Pa

c. 1.78×10^3 torr

5-2. a. 15.8 L

b. 3.17 atm

5-3. a. 52°C = 325 K; 104°C = 377 K; 6.75 L

b. 297 K = 24°C

5-4. a. 12.4 L

b. 10.1 L

c. 1.36 moles

5-5. 0.484 moles of $CaCO_3$ →

0.484 moles of CO_2 at 0.852 atm and 624 K →

29.1 L

5-6. a. 30.1 g/mole

b. C_2H_6

5-7. a. P_{N_2} = 504 mm Hg

b. P_{O_2} = 135 mm Hg

c. P_{Ar} = 6.5 mm Hg

5-8. a. Both tanks contain the same number of molecules of gas; P = nRT/V is the same for both tanks.

b. The temperature is the same in both tanks; the molecules in both tanks have the same average kinetic energy.

c. The H_2 molecules are lighter and travel faster.

5-9. Helium escapes 2.83 times faster than oxygen. $\left[\sqrt{\frac{M_{O_2}}{M_{He}}}\right]$

5-10. a. Attractive forces between molecules reduce the pressure exerted by the molecules hitting the walls of the container.

b. The actual volume occupied by the gas molecules reduces the volume of the container that the molecules can move around in; the gas molecules hit the walls of the container more often. The actual volume of the molecules "increases" the pressure above the "ideal" pressure.

CHAPTER SIX: THERMOCHEMISTRY

This chapter focuses on energy changes which accompany chemical processes.

The Law Of Conservation of Energy states that energy can be transformed, but never created or destroyed.

We will always focus on the system and changes in its energy.

Potential Energy depends on the positions of particles or objects in a system. Potential energy is always measured with respect to a "reference point" where the potential energy is defined as zero.

The choice of the reference point is arbitrary. The choice of the reference point does not affect changes in potential energy, as long as we always use the same reference point.

In the gravitational field of the Earth, the potential energy of an object of mass (m) held at a height (h) above a reference point is:

$$P.E. = m \cdot h \cdot g$$

where g is the acceleration of gravity. An average value for g on the surface of the earth is 9.81 m/s^2.

Problem 6-1:

Consider a ball weighing 0.250 kg held 1.32 m above a table. The top of the table is 0.78 m above the floor.

   a. What is the change in potential energy of the ball when the ball is dropped to the surface of the table?

   If we measure the potential energy with the top of the table as a reference point:

   1.32 m above the table, the ball's potential energy equals:

$$P.E._{initial} = 0.250 \text{ kg} \times 1.32 \text{ m} \times \frac{9.81 \text{ m}}{s^2}$$

$$= \frac{3.23_7 \text{ kg m}^2}{s^2} = \frac{3.24 \text{ kg m}^2}{s^2} = 3.24 \text{ J}$$

   At the surface of the table, h = 0, and the ball's potential energy equals 0. The change in the ball's potential energy is:

$$P.E._{final} - P.E._{initial} = \frac{0 \text{ kg m}^2}{s^2} - \frac{3.24 \text{ kg m}^2}{s^2} = \frac{-3.24 \text{ kg m}^2}{s^2}$$

   The negative sign indicates that the system's potential energy has decreased.

What if we measure the potential energy with the floor as the reference point?

The ball starts 2.10 m (1.32 m + 0.78 m) above the floor, and ends 0.78 m above the floor (on top of the table).

$$P.E._{initial} = 0.250 \text{ kg} \times 2.10 \text{ m} \times \frac{9.81 \text{ m}}{s^2}$$

$$= \frac{5.15 \text{ kg m}^2}{s^2}$$

$$P.E._{final} = 0.250 \text{ kg} \times .78 \text{ m} \times \frac{9.81 \text{ m}}{s^2}$$

$$= \frac{1.91 \text{ kg m}^2}{s^2}$$

$$P.E._{final} - P.E._{initial} = \frac{1.91 \text{ kg m}^2}{s^2} - \frac{5.15 \text{ kg m}^2}{s^2} = \frac{-3.24 \text{ kg m}^2}{s^2}$$

The change in potential energy doesn't depend on the reference point.

b. How fast is this ball travelling just before it hits the surface of the table?

The ball's potential energy has decreased 3.24 kg m²/s².

Therefore, its kinetic energy must have increased 3.24 kg m²/s².

The ball started at rest, 1.32 m above the table, with zero kinetic energy. Its kinetic energy just before it hits the table must be 3.24 kg m²/s². We can find the ball's velocity.

$$K.E. = \frac{1}{2} mu^2$$

$$\frac{3.24 \text{ kg m}^2}{s^2} = \frac{1}{2} (0.250 \text{ kg}) u^2$$

$$u^2 = \frac{2 \cdot 3.24 \text{ kg m}^2}{0.250 \text{ kg s}^2} = \frac{38.8 \text{ m}^2}{s^2}$$

$$u = \frac{6.23_5 \text{ m}}{s} = \frac{6.24 \text{ m}}{s}$$

The SI unit of energy, the Joule (J) equals 1 kg m²/s².

c. What has happened to the ball's kinetic energy when it bounces and then finally comes to rest on the table?

The macroscopic kinetic energy of the ball has been converted into microscopic kinetic energy of the molecules in the ball, the air, and the table, all of which are now warmer.

In an exothermic (heat is released) chemical reaction, the potential energy of the system (reactants and products) decreases, and energy is transferred to the surroundings.

Heat

Although we often speak of heat as though it were a substance, heat is actually a method of transferring energy. Heat is the transfer of energy between two objects, or between the system and its surroundings, that is caused by a temperature difference. Energy always flows from the warmer object to the cooler object. Part of the energy transferred "as heat" can be used to do work, while part of it is "stored" within the system as microscopic kinetic energy, and part of it is stored as potential energy. We can never completely convert all of the energy transferred as heat to work. We'll look again at thermodynamics in Chapter 16.

Work

Work, like heat, is not a substance, but a method of transferring energy. Work can be mechanical, chemical or electrical. When we use a piston to compress a gas, we are doing mechanical work. When we charge a battery, we are doing electrical work. When living cells transport ions and molecules across membranes, they are doing chemical work. We will begin by looking at mechanical work.

Mechanical work is defined as the force exerted on an object multiplied by the distance through which the object moves. If you push very hard on a wall you may feel as though you are "working" hard. However, if the wall doesn't move, you haven't done any mechanical work, by this definition.

Energy

We will always take the system's view for any energy change. The change in energy will be positive when the system's energy increases, and negative when the system's energy decreases. We will use Δ to represent "change in."

Then, $\Delta E = E_{final} - E_{initial}$

The First Law of Thermodynamics

The energy of the universe is constant. Any change in the energy of a system, then, must be due to work done on the system (by the surroundings), or heat transferred to the system (from the surroundings). We'll represent the energy transferred to the system as work by $\underline{w}$, and the energy transferred to the system as heat by $\underline{q}$.

Then, $\Delta E = q + w$

That is, the change in the system's internal energy equals the energy transferred to the system as heat plus the energy transferred to the system as work.

The state of a system can be described by its composition, pressure, volume, and temperature. An equation of state relates these properties. An example of an equation of state is: PV = nRT, for an ideal gas.

The energy of a system is a state function. That is, the energy of the system depends only on the state of the system, not on its history. When a system goes from one state to another, the change in energy depends only on the initial state, and the final state, not on the pathway.

The potential energy of our ball held 1.32 m above the table is a state function. The change in potential energy of the ball initially 1.32 m above the table, and finally resting on the table is -3.24 kg m^2/s^2 and does not depend on the ball's path. You could have dropped the ball, or gently lowered the ball, or bounced the ball off the floor and then placed it on the table. The change in potential energy of the ball depends only on its initial state and its final state.

Heat (q) and work (w) are not state functions. Energy transferred as heat to a system, and energy transferred as work to a system each depend on the path. For example, you could raise the temperature of a beaker of water by heating it, or by stirring the water, or by both heating and stirring.

Calculating Work Done

Suppose we compress a gas with a piston. How much work have we done on the system?

Look at Figure 6.4 in the text:

$$\text{Work} = \text{Force} \times \text{Distance}$$

To simplify our calculations, suppose we push with a constant force F on a piston of area A, and the piston moves the distance Δh.

The work done on the system is:

$$w = F \cdot \Delta h$$

We could express the force as the pressure exerted on the piston times the area of the piston. [This pressure (P_{ext}) is not the pressure exerted by the gas inside the piston.]

$$w = P_{ext} \cdot A \cdot \Delta h$$

The volume of the gas changes by $A \cdot \Delta h$. Actually $\Delta V = V_{final} - V_{initial} = -A \cdot \Delta h$. When we compress the gas, we do work on the system and the volume decreases.

$$\text{Then, } w = -P_{ext} \Delta V$$

The units for work depend on the units of pressure and volume. If we measured pressure in Pascals (Pa) and volume in cubic meters (m^3), we would calculate work in Joules. If we measure pressure in atmospheres (atm) and volume in liters (L) we calculate work in liter • atm. We can convert this work to Joules using the factor:

$$1 \text{ L} \cdot \text{atm} = 101.3 \text{ J}$$

Problem 6-2:

a. How much work is done on a gas when 3.82 L are compressed to 1.67 L under a pressure of 2.65 atm?

$$w = -P_{ext} \, \Delta V$$

$$= -P_{ext} \, (V_{final} - V_{initial})$$

V decreases, therefore ΔV is negative, and the work done on the system is positive.

$$w = -2.65 \text{ atm} \, (1.67 \text{ L} - 3.82 \text{ L})$$

$$= -2.65 \text{ atm} \, (-2.15 \text{ L}) = 5.69_7 \text{ L} \cdot \text{atm}$$

Converting to Joules:

$$w = 5.69_7 \text{ L} \cdot \text{atm} \times \frac{101.3 \text{ J}}{\text{L} \cdot \text{atm}} = 577._1 \text{ J}$$

$$= 577 \text{ J}$$

b. If this gas is compressed and 159 J are added as heat to the gas, what is the total change in energy of the gas?

$$\Delta E = q + w$$

$$= 159 \text{ J} + 577 \text{ J} = 736 \text{ J}$$

c. If the piston is suddenly moved back to its original position, the gas expands to fill the space. How much work did the gas do when expanding?

$$w = -P_{ext} \, \Delta V$$

If the gas expanded "against no pressure," no work was done when the gas expanded.

d. If the gas is heated, and the pressure on the piston increases, but the piston doesn't move, how much work does the gas do?

$$w = -P_{ext} \, \Delta V$$

If the volume doesn't change, that is, if the piston doesn't move, no work is done.

Calculating Heat Added

When we add "heat" to a substance, the temperature changes. The amount of heat required to raise the temperature is proportional to the change in temperature.

$$q = C \cdot \Delta T$$

where q is the heat added and, ΔT is the change in temperature. The proportionality constant C is the heat capacity in J/°C.

An object with a large heat capacity absorbs more heat when it is warmed one degree than an object with a small heat capacity.

The amount of heat required to raise the temperature of a substance is also proportional to the amount of the substance. When the size of the sample is measured in grams, we can use the heat capacity per gram C_m (in J/°C g) and the mass, m:

$$\text{Then:} \quad q = C_m \cdot m \cdot \Delta T$$

When the size of the sample is measured in moles, we can use the molar heat capacity, the heat capacity per mole, C_n (in J/°C mol) and the number of moles, n:

$$\text{Then:} \quad q = C_n \cdot n \cdot \Delta T$$

The heat capacity of a substance depends on the pathway for the process. Chemists use constant volume heat capacities and constant pressure heat capacities. For liquids and solids, there is little difference between these two heat capacities. For gases, the heat capacity at constant pressure is larger than the heat capacity at constant volume. When we add heat to a calorimeter we can't use C_m or C_n for the calorimeter because the calorimeter is made of many different substances. We can use the heat capacity of the <u>whole</u> calorimeter.

Calorimetry

In principle, a calorimeter traps all the heat produced by a reaction. We can calculate the heat produced by measuring the temperature change, and using the heat capacities of the contents and the container.

Problem 6-3:

A 12.5 g metal block was heated to 100.0°C and dropped into a calorimeter containing 48.0 g of water at 24.37°C. After the block was dropped into the water, the temperature of the water rose to 26.43°C. Assuming that the calorimeter itself absorbs a negligible amount of heat:

a. How much heat is lost by the block?

Solution:

If all the heat is trapped in the calorimeter:

The heat lost by the metal block equals the heat gained by the water.

The heat gained by the water, using the heat capacity from Table 6.1.

$$q = C_m \cdot m \cdot \Delta T$$

$$= \frac{4.18 \text{ J}}{^{\circ}C \text{ g}} \cdot 48.0 \text{ g} \cdot (26.43 - 24.37^{\circ}C)$$

$$= \frac{4.18 \text{ J}}{^{\circ}C \text{ g}} \cdot 48.0 \text{ g} \cdot 2.06^{\circ}C = 413._3 \text{ J}$$

b. What is the heat capacity of the block? All the heat lost by the block equals the heat gained by the water, 26.43°C.

Solution:

$$q = C \cdot \Delta T$$

$$C = \frac{q}{\Delta T}$$

$$413 \text{ J} = C \ (100.0^{\circ}C - 26.43^{\circ}C)$$

The block ends up at the same temperature as the water.

$$C = \frac{413 \text{ J}}{73.6^{\circ}C} = \frac{5.61 \text{ J}}{^{\circ}C}$$

c. What is the heat capacity per gram of the metal?

Solution:

$$C_m = \frac{\dfrac{5.61 \text{ J}}{^{\circ}C}}{12.5 \text{ g}} = \frac{.449 \text{ J}}{^{\circ}C \text{ g}}$$

We're interested in using calorimeters to measure the heat released or absorbed when a reaction takes place.

Problem 6-4:

25.00 mL of 0.200 M $AgNO_3$ is added to 35.00 mL of 0.200 M NaCl. AgCl precipitates and the temperature of the mixture rises 1.30°C. Given that the heat capacity of the solution is 4.20 J/°C g, and the density of the solution is 1.000 g/mL:

a. How much heat was released by the reaction $Ag^+(aq) + Cl^-(aq) \rightarrow AgCl$?

$$q = C_m \cdot m \cdot \Delta T$$

$$= \frac{4.20 \text{ J}}{\degree C \text{ g}} \cdot 60.00 \text{ g} \cdot 1.30\degree C$$

$$= 327._6 \text{ J} = 328 \text{ J}$$

b. How many moles of silver chloride formed?

$$\text{moles of } Ag^+ = \frac{0.200 \text{ moles}}{\text{liter}} \times .02500 \text{ L} = 5.00 \times 10^{-3} \text{ moles}$$

$$\text{moles of } Cl^- = \frac{0.200 \text{ moles}}{\text{liter}} \times .03500 \text{ L} = 7.00 \times 10^{-3} \text{ moles}$$

Ag^+ is the limiting reagent.

Therefore, 5.00×10^{-3} moles of AgCl formed.

c. How much heat would have been produced if one mole of silver chloride had formed?

$$\frac{328 \text{ J}}{5.00 \times 10^{-3} \text{ moles AgCl}} = 6.56 \times 10^4 \text{ J/mole}$$

$$= 65.6 \text{ kJ/mole}$$

65.6 kJ is the "heat of reaction" for the reaction:

$$Ag^+(aq) + Cl^-(aq) \rightarrow AgCl(s)$$

The heat of reaction depends on the conditions under which we carry out the reaction. If we carry out the reaction in a "constant volume" calorimeter, so that the volume doesn't change, no PV work takes place, and $\Delta E = q$. When we measure the heat of reaction at constant volume we are measuring ΔE. However, many reactions are studied in a "constant pressure" calorimeter, where there <u>is</u> a change in the volume.

Then:
$$\Delta E = q + w = q - P\Delta V$$

Under constant pressure, then

$$q = \Delta E + P\Delta V$$

If we had a new property $H = E + PV$, then, when we measure the heat of reaction under constant pressure,

$$q = \Delta E + P\Delta V = \Delta H$$

we are measuring ΔH. This new property, H, is called <u>enthalpy</u>. Enthalpy is useful when we study reactions at constant pressure. Actually, for reactions involving only liquids, solutions, and solids, ΔE and ΔH are very nearly the same. For reactions involving gases, ΔE and ΔH may be very different.

When a process absorbs heat, q and ΔH are both positive. When a process produces heat, q and ΔH are negative. Remember, we take the system's point of view.

For our reaction:

$$Ag^+(aq) + Cl^-(aq) \rightarrow AgCl(s)$$

the "system" is the mixture of reactants. Heat is given off to the surroundings. q = -65.6 kJ for each mole of AgCl formed.

ΔH for the reaction is -65.6 kJ.

$$Ag^+(aq) + Cl^-(aq) \rightarrow AgCl(s) \quad \Delta H_{rxn} = -65.6 \text{ kJ}$$

The "heat of reaction" (or enthalpy of reaction) ΔH is the change in enthalpy for the reaction as written.

For the reaction:

$$2 \ C_2H_6(g) + 7 \ O_2(g) \rightarrow 4 \ CO_2(g) + 6 \ H_2O(l)$$

$$\Delta H = -3.12 \times 10^6 \text{ J}$$

That is:

3.12×10^6 J are produced when 2 moles of ethane (C_2H_6) react with 7 moles of oxygen gas to form 4 moles of carbon dioxide and 6 moles of water.

Problem 6-5:

How many Joules are produced when 18.2 g of ethane (C_2H_6) are burned in excess oxygen?

Solution: $\text{g } C_2H_6 \xrightarrow{MW} \text{moles } C_2H_6 \xrightarrow{\text{bal'd eqn}} \text{heat evolved}$

$$\text{moles } C_2H_6 = \frac{18.2 \text{ g}}{\dfrac{30.07 \text{ g}}{\text{mole}}} = .605 \text{ moles } C_2H_6$$

The balanced equation and ΔH indicate that 3.12×10^6 J are produced when 2 moles of ethane are burned.

$$\text{Heat released} = \frac{3.12 \times 10^6 \text{ J}}{2 \text{ moles } C_2H_6} \times .605 \text{ moles } C_2H_6$$

$$= 9.44 \times 10^5 \text{ J}$$

Now we can expand our "stoichiometry map" even further to include energy lost or gained.

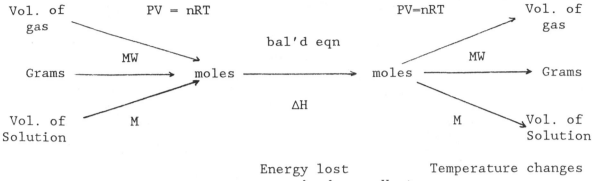

Hess's Law

Enthalpy is a state function, and the change in enthalpy in going from one state to another doesn't depend on the pathway. This principle allows us to calculate the heat of reaction for a process we can't carry out very easily.

If we can write an overall reaction as the sum of a series of steps, the ΔH for the overall reaction is the sum of the ΔH's for each of the steps.

Problem 6-6:

Given ΔH_{rxn} for these two reactions,

i) $C(gr) + O_2(g) \rightarrow CO_2(g)$ $\Delta H = -393.51$ kJ

and

ii) $2\ CO(g) + O_2(g) \rightarrow 2\ CO_2(g)$ $\Delta H = -565.68$ kJ

Calculate ΔH_{rxn} for the overall reaction:

$$2\ C(gr) + O_2(g) \rightarrow 2\ CO(g)$$

Solution:

a.   Combine the reactions:

To get the overall reaction, we need to start with two moles of carbon: then multiply equation i) by two to yield two moles of carbon dioxide.

2x i): $2\ C(gr) + 2\ O_2(g) \rightarrow 2\ CO_2(g)$ $\Delta H = 2(-393.51$ kJ$)$

$= -787.02$ kJ

We can then reverse reaction ii), to form carbon monoxide from carbon dioxide.

$$-ii) \quad 2\ CO_2(g) \rightarrow O_2(g) + 2\ CO(g) \qquad \Delta H = -(-565.68\ kJ)$$

$$= 565.68\ kJ$$

Add these reactions:

$$2\ C(gr) + 2\ O_2(g) + 2\ CO_2(g) \rightarrow 2\ CO_2(g) + O_2(g) + 2\ CO(g)$$

Subtract molecules which appear on both sides:

$$2\ C(gr) + 2\ O_2(g) + 2\ CO_2(g) \rightarrow 2\ CO_2(g) + O_2(g) + 2\ CO(g)$$
$$1$$

$$2\ C(gr) + O_2(g) \rightarrow 2\ CO(g)$$

Add the ΔH's for the two reaction we added:

$$\Delta H = -787.02\ kJ + 565.68\ kJ$$

$$= -221.34\ kJ$$

Enthaply of Formation

Hess's Law allows us to use "enthalpies of formation" for compounds to calculate ΔH_{rxn} for reactions. The enthalpy of formation of a compound is the change in enthalpy when one mole of a compound in its "standard state" is formed from its constituent elements (in their standard states).

The enthalpy of formation, ΔH_f°, of ammonia gas refers to the reaction:

$$\frac{1}{2}\ N_2(g) + \frac{3}{2}\ H_2(g) \rightarrow NH_3(g) \qquad \Delta H_f^\circ = -46.19\ kJ/mol$$

Enthalpies of formation are always given on a per mole basis. To calculate ΔH for a reaction, we imagine carrying out the reaction, first breaking the reactants down into the elements and then forming the products from the elements.

$$\Delta H_{rxn} = \Sigma \Delta H_{f_{products}} - \Sigma \Delta H_{f_{reactants}}$$

Problem 6-7:

Calculate ΔH_{rxn} for:

$$2\ CO(g)\ +\ O_2(g)\ \rightarrow\ 2\ CO_2(g)$$

from the enthalpies of formation of $CO(g)$ and $CO_2(g)$.

Substance	ΔH_f
$CO_2(g)$	−393.51 kJ
$CO(g)$	−110.67 kJ

Solution:

$$\Delta H\ =\ 2\ \Delta H_{f_{CO_2}}\ -\ \Delta H_{f_{O_2}}\ -\ 2\ \Delta H_{f_{CO}}$$

What is the enthalpy of formation of $O_2(g)$? The enthalpy of formation of an element in its standard state is exactly zero!

Then:
$$\Delta H_{rxn}\ =\ 2(-393.51)\ -\ (0)\ -\ 2(-110.67)\ kJ$$

$$=\ -787.02\ +\ 221.34\ kJ$$

$$=\ -565.68\ kJ$$

CHAPTER SIX: SELF-TEST

6-1. The specific heat of a metal is 0.382 J/g°C.

How much heat is required to raise the temperature of a 35.0 g block of this metal 12.6°C?

6-2. a. A system <u>gains</u> 247 J of heat and <u>does</u> 489 J of work; calculate ΔE.

b. The same system <u>loses</u> 315 J of heat and has 73 J of work done on it; calculate ΔE.

6-3. A 50.0 g sample of $CaCO_3$ is heated to completely drive off CO_2:

$$CaCO_3(s) \xrightarrow{\Delta} CaO(s) + CO_2(g)$$

a. What volume of CO_2 is produced at 250°C and 1.25 atm?

b. How much work is done <u>by</u> the system if this process is carried out in a container with a sliding piston?

6-4. A 45.0 g of a block of metal at 99.5°C is dropped into 150.0 g of water at 24.00°C. The temperature of the water rises to 26.40°C. The specific heat of water is 4.18 J/g°C.

a. How much "heat" is gained by the water?

b. How much "heat" is lost by the block of metal?

c. What is the specific heat of the metal?

6-5. The reaction for the combustion of acetylene is:

$$2\ C_2H_2(g) + SO_2(g) \rightarrow 4\ CO_2(g) + 2\ H_2O(g) \qquad \Delta H = -2511\ kJ$$

How much heat is evolved when 35.58 g of C_2H_2 is burned in excess oxygen?

6-6. Calculate ΔH for the addition of hydrogen gas to ethylene to form ethane:

$$C_2H_4(g) + H_2(g) \rightarrow C_2H_6(g)$$

Using these data:

i) $C_2H_6(g) + \frac{7}{2} O_2(g) \rightarrow 2\ CO_2(g) + 3\ H_2O(g)$ $\Delta H_1 = -1323$ kJ

ii) $C_2H_4(g) + 3\ O_2(g) \rightarrow 2\ CO_2(g) + 2\ H_2O(g)$ $\Delta H_2 = -1429$ kJ

and iii) $H_2(g) + \frac{1}{2} O_2(g) \rightarrow H_2O(g)$ $\Delta H_3 = -242$ kJ

(Hint: How can you combine these reactions to give the desired reaction?)

6-7. Use enthalpies of formation to calculate ΔH for the reaction:

$$C_2H_5OH(\ell) + 3\ O_2(g) \rightarrow 2\ CO_2(g) + 3\ H_2O(g)$$

Substance	ΔH_f°
$C_2H_5OH(\ell)$	-277.7 kJ/mole
$CO_2(g)$	-393.5 kJ/mole
$H_2O(g)$	-241.8 kJ/mole
$O_2(g)$	0.0 kJ/mole

CHAPTER SIX: SELF-TEST ANSWERS

6-1.   168 J

6-2. $E = q + w$

    a. $E = 247\ J - 489\ J = -242\ J$

    b. $E = -315\ J + 73\ J = -242\ J$

6-3.   0.500 moles $CaCO_3 \rightarrow$ 0.500 moles CO_2

    a. $V = \dfrac{nRT}{P} = \dfrac{0.500\ moles\ \dfrac{(0.08206\ L \cdot atm)}{mole}\ (523\ K)}{1.25\ atm} = 17.2\ L$

    b.   Work done <u>on</u> the system: $W - -P_{ext} \cdot \Delta V$

$$= -1.25\ atm\ (17.2\ L)$$

$$= -21.5\ L \cdot atm$$

Work done <u>by</u> the system: $+ 21.5\ L \cdot atm$

$$1\ L \cdot atm = 101.3\ J$$

Work done by the system $= 21.5\ L \cdot atm \times \dfrac{101.3\ J}{L \cdot atm} = 2.2 \times 10^3\ J$

6-4.   a. $1.50 \times 10^3\ J$

    b. $1.50 \times 10^3\ J$

    c. $0.456\ J/g°C = \dfrac{1.50 \times 10^3\ J}{45.0\ g\ (99.5 - 26.4\ °C)}$

6-5.   1.366 moles C_2H_2 are burned

$$1715\ kJ = 1.366\ moles \times \dfrac{2511\ kJ}{2\ moles}$$

6-6. $\Delta H = \Delta H_2 - \Delta H_1 + \Delta H_3$

$$= -348\ kJ$$

6-7. $\Delta H = 2\ \Delta H°_{f(CO_2)} + 3\ \Delta H°_{f(H_2O)} - \Delta H°_{f(C_2H_5OH)} - 3\ \Delta H°_{f(O_2)}$

$$= -1235\ kJ$$

CHAPTER SEVEN: THE PERIODIC TABLE AND ATOMIC STRUCTURE

We first looked at atoms and ions in Chapter 2. Now, in Chapter 7, we take a deeper look at the internal structure of the atom. We will focus on models describing electrons in atoms.

Waves

Consider an ocean wave far from the shore: if we watch a cork in the water, it rides up and down as the wave passes, but the cork does not travel forward. The wave is a disturbance travelling through the water, but the water itself doesn't travel. We could determine the frequency of the wave by counting how many times the cork rose up during a certain time period, or by counting how many crests passed a given point in a certain time period. The frequency is given in cycles (number of waves) per second. As the "number-of-waves" has no units, frequency has the units $1/s$, or s^{-1}. The wavelength of a wave is the distance between two consecutive peaks or troughs, and has units of length: meters, centimeters, nanometers, etc.

Problem 7-1:

A pebble is dropped in a pond. The distance between peaks for the waves in the water is 1.75 cm. Five peaks pass a given point in one second. How fast are the waves travelling?

The frequency of the wave is $5 \ s^{-1}$. The wavelength is 1.75 cm.

Solution:

The wave has travelled 5×1.75 cm in one second:

The wave's velocity is $\dfrac{5 \times 1.75 \ cm}{s} = \dfrac{8.75 \ cm}{s}$

For any wave: velocity = wavelength × frequency

$$\upsilon = \lambda \cdot \nu$$

Electromagnetic Radiation

All electromagnetic radiation (light, radiowaves, microwaves, x-rays) travels at the same velocity in a given medium. Electromagnetic radiation is characterized by its frequency and its wavelength. Water waves are associated with "oscillations" of the surface of water. Electromagnetic radiation is associated with "oscillating" electric and magnetic fields.

Quantum Theory

The results of quantum theory may challenge your intuition about the nature of matter. The challenges include these results:

i) Energy is quantized; that is, only certain values of energy are allowed. As only certain values of energy are allowed, only certain energy changes are allowed. When an atom or molecule emits "light" (electromagnetic radiation), the frequency of the light allows us to calculate the change in energy of the atom or molecule:

$$\Delta E = h\nu$$

where h is Planck's constant, $6.6262 \times 10^{-34} \frac{kg \ m^2}{s^2}$

ii) Light can behave as a wave or as a stream of particles (photons). The energy associated with a photon of light can be calculated from the frequency or the wavelength of the light:

$$E_{photon} = h\nu = \frac{hc}{\lambda}$$

Light with a short wavelength (higher frequency), such as violet light has a higher energy than light with a long wavelength, such as red light.

iii) Every moving particle has a wave associated with it. That is every "object" can behave as a particle or a wave, depending on the circumstances. The wavelength associated with a moving particle can be calculated:

$$\lambda = \frac{h}{mv}$$

If all matter can act as a particle or as a wave, why don't we see baseballs behaving as waves? When do "particles" behave as waves? When are interactions "wave-like"? When are they "particle-like"?

As a qualitative rule, when two objects interact, if the wave-length of the moving object is small compared to the size of the other object, their interaction will be "particle-like." The 100 gm ball moving at 35 m/s (in sample exercise 7.3) has a wavelength of 1.9×10^{-34} m. This wavelength is very small compared to the size of a bat; therefore, the interaction between bat and ball is "particle-like."

If the wave-length of the moving object is about the same magnitude as the size of the other object, or larger than the size of the object, the interaction is "wave-like." Therefore, a beam of electrons can be diffracted by the atoms in a solid.

95

Spectra

A spectrum is a graph of the intensity of the radiation, emitted by a sample or absorbed by a sample, plotted against the wavelength (or frequency) of the radiation. We can see the spectrum of visible light by passing the ray of light through a prism. The prism bends the path of light with shorter wavelengths more than it bends the path of light with longer wavelengths.

When a ray of light (visible electromagnetic radiation) from a hot, glowing source is passed through a prism, the prism separates the light of different wavelengths into a continuous spectrum, a "rainbow" with no wavelengths missing.

When a sample of hydrogen gas is placed between two electrodes in a sealed tube, and a high voltage difference is applied to the electrodes, the tube glows. When a ray of this light is passed through the prism, we see a series of brightly colored separate lines, a line spectrum.

This series of lines is characteristic of hydrogen. Each element has its own characteristic line spectrum. The visible line spectrum of an element give us information about the electrons in the atoms of the element. We can calculate the energy of the light for each of these lines.

Problem 7-2:

The four visible lines in the hydrogen gas emission spectrum have the following wavelengths. Calculate the energy corresponding to each line:

$$656.3 \text{ nm}, \quad 486.3 \text{ nm}, \quad 434.2 \text{ nm}, \quad 410.1 \text{ nm}$$
$$\text{red} \qquad \text{green} \qquad \text{blue violet} \qquad \text{violet}$$

Solution:

Use the equation:

$$E = \frac{hc}{\lambda}$$

(Remember, a nanometer [nm] is 1.10^{-9} meters [m].)

h: Planck's constant = 6.6262×10^{-34} J $\times$ s

c: speed of light = 2.9979×10^{8} m/s

For the line with wavelength 656.3 nm (656.3×10^{-9} m):

$$E = \frac{6.6262 \times 10^{-34} \text{ J} \times \text{s}}{656.3 \times 10^{-9} \text{ m}} \times \frac{2.9979 \times 10^{8} \text{ m}}{\text{s}}$$

$$= 3.026_7 \times 10^{-19} \text{ J} = 3.027 \times 10^{-19} \text{ J}$$

96

Similarly:

For the line of wavelength 486.3 nm: $E = 4.085 \times 10^{-19}$ J

For the line of wavelength 434.2 nm: $E = 4.575 \times 10^{-19}$ J

For the line of wavelength 410.1 nm: $E = 4.844 \times 10^{-19}$ J

Where does the energy of the emitted light come from? The electrical discharge causes some of the hydrogen molecules to dissociate. The resulting hydrogen atoms are "excited" and emit light as they lose their excess energy.

We observe that only certain wavelengths of light are emitted. Each wavelength corresponds to a particular energy. Therefore, only certain changes in energy of the atom are allowed. The simplest explanation of this observation is that only certain energies of the atom are allowed, that is, the energy of the atom is quantized.

Angular Momentum

We can think of momentum as a measure of the tendency for an object to keep moving. The momentum of an object travelling in a straight line is the mass of the object multiplied by its velocity, the linear momentum:

$$momentum = mv$$

The angular momentum for an object travelling in a circle (for example, a ball whirled on the end of a string) is the mass of the object multiplied by its velocity and multiplied by the radius of the circular path.

$$angular\ momentum = mvr$$

The Bohr Model of the Atom

Niels Bohr proposed a model for the atom with electrons moving in circular orbits about the nucleus. He assumed that only certain orbits were allowed, those where the angular momentum of the electron was a multiple of $h/2\pi$.

Classical physics predicts that if a charged particle, like the electron, travelled in a circular path around a second charged particle of opposite sign, like the nucleus, the first particle would spiral in towards the second, constantly emitting electromagnetic radiation. Bohr's model assumed that this classical behavior didn't happen. The only justification for Bohr's assumption was that his model did predict the allowed energy levels for an electron in the hydrogen atom and his model did predict the observed wavelengths in the hydrogen emission spectrum. Energy (light) would be emitted when an electron "jumped" from a higher energy orbit to a lower energy orbit.

Bohr's model predicted the allowed energies for an electron in a hydrogen atom to be:

$$E = \frac{-2\pi^2 me^4 Z^2}{h^2 n^2}$$ (Equation 7.1)

where n can be any positive integer: 1, 2, 3, 4....

Problem 7-3:

What is the lowest energy an electron can have in the hydrogen atom?

$$E = -\left[\frac{2\pi^2 me^4}{h^2} \quad \frac{Z^2}{n^2}\right]$$

Solution:

As n increases, the term within the brackets becomes smaller in magnitude, and E becomes a smaller negative number. Therefore, as n increases, E approaches 0 (zero). The lowest energy, the most negative energy then, must be for n = 1.

The larger n is for an orbit, the larger the orbit radius and the more positive (less negative) the energy of an electron in the orbit. This agrees with our intuition. It must take work to move an electron farther from the positively charged nucleus. The potential energy of the electron increases (becomes more positive, less negative) as we move the electron away from the nucleus.

Problem 7-4:

a. Calculate the energy of the photon emitted when an electron "jumps" from a Bohr orbit with n = 3 to a Bohr orbit with n = 2.

$$E_n = -2.178 \times 10^{-18} \text{ J} \times \frac{1}{n^2}$$ (Equation 7.2)

$$E_3 = -2.178 \times 10^{-18} \text{ J} \left(\frac{1}{3^2}\right)$$

$$= -2.420 \times 10^{-19} \text{ J}$$

$$E_2 = -2.178 \times 10^{-18} \text{ J} \left(\frac{1}{2^2}\right)$$

$$= -5.445 \times 10^{-19} \text{ J}$$

$$\Delta E_{electron} = E_{final} - E_{initial}$$

$$= E_2 - E_3 = (-5.445 \times 10^{-19} \text{ J}) - (-2.420 \times 10^{-19})$$

$$= -3.025 \times 10^{-19} \text{ J}$$

The energy of the electron decreases. But energy is conserved. Therefore, the photon of light must have an energy of 3.025×10^{-19} J.

b. What is the wavelength of light emitted when an electron "jumps" from the Bohr orbit with n = 3 to that with n = 2?

$$E_{photon} = \frac{hc}{\lambda}$$

$$\lambda_{wavelength} = \frac{hc}{E} = \frac{6.6262 \times 10^{-34} \text{ J s}}{3.025 \times 10^{-19} \text{ J}} \times \frac{2.9979 \times 10^{8} \text{ m}}{s}$$

$$= 6.567 \times 10^{-7} \text{ m}$$

1 m = 10^9 nm: $\quad$ = 656.7 nm

Look back at problem 7.2 in this chapter. The 3 → 2 transition corresponds to the longest wavelength line in the visible hydrogen emission spectrum.

Although Bohr's model worked well for the hydrogen atom, and other one-electron "atoms" (He^+, Li^{2+}), it didn't work well for multi-electron atoms.

Wave Mechanics

Schrödinger built on De Broglie's idea that all particles have a wavelength, and Bohr's model of the hydrogen atom with quantized energy and quantized angular momentum for the electron. Schrödinger applied classical equations for waves to the problem of an electron in an atom.

The Schrödinger equation is a differential equation, that is, an equation that involves derivatives. We are looking for functions that satisfy the equation:

$$H\psi = E\psi$$

H is an operator, a set of mathematical instructions (3 is an operator, indicating "multiply what follows by 3"; $\sqrt{}$ is an operator, indicating "take the square root of what follows").

E is a constant, which turns out to be the energy of the atom.

Schrödinger's equation says: "find all the functions that H will operate on to yield the same function times a constant."

There is more than one function that satisfies the Schrödinger equation. When you solve a quadratic equation:

$$ax^2 + bx + c = 0$$

to find values of x that satisfy the equation, there are <u>two</u> values of x:

$$x = \frac{-b + \sqrt{b^2 - 4\ ac}}{2\ a} \quad \text{and} \quad x = \frac{-b - \sqrt{b^2 - 4\ ac}}{2\ a}$$

Solutions to the Schrödinger equation, are <u>equations</u>, not values, but the principle is the same. You do not need to be <u>able</u> to solve the equation to use the solutions.

<u>Solutions to the Schrödinger Equation</u> (An optional look at the "terrifying" wave functions.) What do the solutions to the Schrödinger equation look like?

First, only certain energies are allowed:

$$E = \frac{-2\pi^2 mc^4}{h^2} \ \frac{Z^2}{n^2} \ ,$$

The same energies predicted by the Bohr model.

Second, the solutions, "the wave functions" contain integers, quantum numbers, n, ℓ, and m_ℓ.

We represent these functions by ψ. ψ depends on position. We often use polar coordinates for a point: r, the distance of the point from the nucleus; and two angles, θ and ϕ. θ, the angle a line from the nucleus to the point makes with the z axis; and ϕ an angle in the x, y plane.

For $n = 1$, $\ell = 0$ and $m = 0$:

The wave function is found to be:

$$\psi_{n\,\ell m} = \psi_{100} = \frac{1}{\sqrt{\pi}} \left(\frac{Z}{a_0}\right)^{3/2} e^{\frac{-Zr}{2a_0}}$$

a_0 is a constant: $\quad a_0 = \frac{h^2}{4\pi^2 me^2}$

Z is the charge on the nucleus

r is the distance from the nucleus

and e is a constant, $2.7182818\ldots$.

Don't panic. You won't use this equation to do calculations. Let's take a qualitative look at how ψ_{100} behaves. As the distance from the nucleus r increases, ψ_{100} becomes exponentially smaller, approaching zero at large distances from the nucleus. Does the direction from the nucleus matter? θ and ϕ don't appear in this equation, so ψ_{100} doesn't depend on direction from the nucleus.

ψ_{100} is the 1s orbital.

For n = 2, ℓ = 0 and m = 0:

$$\psi_{200} = \frac{1}{4\sqrt{2\pi}} \left(\frac{Z}{a_0}\right)^{3/2} \left(2-\frac{Zr}{a_0}\right) e^{\frac{-Zr}{2a_0}}$$

Again, how does ψ_{200} behave? Ignore the constants in front. This wave function doesn't depend on direction from the nucleus, either. The part of ψ_{200} in brackets $(2 - Zr/a_0)$ equals 2 when r = 0; decreases as r increases; equals zero when Zr/a_0 = 2; and then becomes negative. The term

$$e^{\frac{-Zr}{2a_0}}$$

just decreases

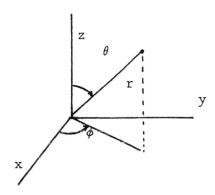

exponentially as r increases. The important points about this wave function are:

ψ_{200} doesn't depend on direction

ψ_{200} has a node, that is, ψ = 0, at a certain distance from the nucleus and ψ_{200} approaches 0 (zero) at large distances from the nucleus.

ψ_{200} is the 2s orbital.

For n = 2, ℓ = 1 and m = 0: The wave function is even more complicated.

$$\psi_{210} = \frac{1}{4\sqrt{2\pi}} \left(\frac{Z}{a_0}\right)^{3/2} \frac{Zr}{a_0} e^{\frac{-Zr}{2a_0}} \cos\theta$$

Ignore the constants in front: look at these parts:

$\dfrac{Zr}{a_0}$ is zero at the nucleus and increases as r increases.

$e^{\frac{-Zr}{2a_0}}$ equals 1 at the nucleus and decreases as r increases

The product:

$$\frac{Zr}{a_0} \times e^{\frac{-Zr}{2a_0}}$$

increases at first, and then decreases, as r increases.

ψ_{210} depends on θ: $\cos \theta = 1$ when $\theta = 0°$, or $180°$, that is, when the point lies on the z-axis; $\cos \theta = 0$ when $\theta = 90°$, that is when the point lies in the xy plane. ψ_{210} will be largest along the z-axis, and has a node in the xy plane. ψ_{210} doesn't depend on direction <u>away</u> <u>from</u> <u>the</u> <u>z-axis</u>. This wave function is "cylindrically symmetric."

ψ_{210} is the pz orbital.

Solutions to the Schrödinger Equation (A summary)

From this short look at wavefunctions we find that some solutions to the wave equation depend only on the distance from the nucleus, while others depend on direction as well as the distance from the nucleus. Some solutions to the wave equation have nodes, that is surfaces where the wave function equals zero.

Wave functions with the same ℓ all have the same general shape. s orbitals ($\ell = 0$) are all spherically symmetric. p orbitals ($\ell = 1$) are cylindrically symmetric. An orbital of larger n is "larger" than an orbital of smaller n.

What does a wave function mean? If ψ^2, the square of the wave function, is a relative probability of finding the electron in a particular place, how can we visualize it? The Bohr model, in many ways, feels more comfortable to us than a probabilistic model; we can more easily imagine electrons travelling in orbits around the nucleus like planets travelling about a sun. With the wave mechanical model, we can never know where the electron <u>is</u> and how it travels from place to place.

Heisenberg's Uncertainty Principle

Heisenberg's Uncertainty Principle states that there is a fundamental limitation on how well we can know the position and momentum of a particle.

The uncertainty in the momentum, of the particle, $\Delta(mv)$, multiplied by the uncertainty in its position, Δx, must be at least $h/4\pi$, where h is Planck's constant.

For large objects, the uncertainty in momentum and velocity is very small. For objects with small masses, the uncertainty in momentum and velocity is so large we must abandon any hope of knowing the exact motion of an electron.

Consider a "thought experiment." Suppose you wish to "see," that is, to locate an electron. Shine a beam of "light" at the electron. If you use "light" (electromagnetic radiation) of a wavelength much longer than the size of the electron, the light wave won't be disturbed by the tiny electron, and will pass by. You will "see" nothing. To "see" the electron we will need to use light whose wavelength is much shorter than the size of the electron. Now the light will bounce back, allowing us to "see" the electron. However, this short wavelength, high energy light acts like a particle hitting the electron, knocking it away from where it was. So, we can see where the electron was, but not where it is, or where it is going.

So we are forced to use a probabilistic model, with orbitals describing the probability of finding an electron in a particular region of space.

Orbitals

In the quantum mechanical description of an atom, electrons are described as "occupying" orbitals, not travelling in orbits. We can represent an orbital in a number of different ways:

i)   An "electron density map," a three-dimensional plot of dots, where the density of the dots in a particular region is proportional to the probability of finding the electron there.

ii) A graph of probability plotted as a function of distance from the nucleus.

iii) A contour diagram, a topographic map of lines of equal probability.

iv) As a "boundary surface," a surface connecting points of equal probability, which contains 90% of the total probability of finding the electron.

Each method has its own advantages.

Quantum Numbers

Each orbital (wave function) is characterized by three quantum numbers: n, ℓ, m:

n, the principle quantum number, describes the size and energy of an orbital.

ℓ, the azimuthal quantum number, describes the shape of an orbital.

m_ℓ, the magnetic quantum number indicates the orientation of an orbital in space.

When we solve the Schrödinger equation, we find one solution to the wave equation for n = 1, ℓ = 0, and m_ℓ = 0. When n = 2, we find one wave function for ℓ = 0 and m_ℓ = 0 and three wave functions for ℓ = 1: m_ℓ = −1, 0, +1.

In fact, for any value of n, we find that ℓ can be any integer from 0 through n-1, and for any ℓ, m_ℓ can be any integer from -1 to +1. For each combination of n, ℓ and m_ℓ there is a different wave function (orbital).

Problem 7-5:

How many different wave functions in an atom can have n = 3?

Solution:

n = 3, ℓ can be: 0, 1, or 2

If ℓ = 0, m_ℓ = 0

If ℓ = 1, m_ℓ can be -1, 0, +1

If ℓ = 2, m_ℓ can be -2, -1, 0, +1, +2

There are <u>nine</u> orbitals with n = 3:

$\psi_{n\,l\,m}$: ψ_{300}, ψ_{31-1}, ψ_{310}, ψ_{311}, ψ_{32-2},

ψ_{32-1}, ψ_{320}, ψ_{321}, ψ_{32-2}

Orbital Shapes

Every orbital can be described by its quantum numbers: n, ℓ, and m_ℓ. Orbitals are commonly described by the principal quantum number n, and a letter to designate ℓ:

ℓ = 0   s orbital

ℓ = 1   p orbital

ℓ = 2   d orbital

ℓ = 3   f orbital

ℓ = 4   g orbital

For an electron in a 4f orbital, n = 4 and ℓ = 3.

All s orbitals are spherically symmetric. s orbitals with n greater than 1 have spherical <u>nodes</u>. Look at figure 7.14 in the text.

All p orbitals are cylindrically symmetric. There are three equivalent p orbitals, (for ℓ = 1, there are three possible values of m_ℓ: -1, 0 + 1) each lying along a different axis. The pz orbital lies along the z-axis, with a node in the xy plane. Look at figure 7.15 in the text.

d orbitals and f orbitals are even more complex. There are five d orbitals (for ℓ = 2, there are five possible values of m_ℓ: -2, -1, 0, +1, +2) and seven f orbitals (for ℓ = 3, seven possible values of m_ℓ: -3, -2, -1, 0, +1, +2, +3).

Nodes are surfaces where there is zero probability of finding an electron.

Spin

In certain experiments, electrons <u>behave</u> as though they are charges spinning like tops. There is a quantum number called the spin, quantum number m_s, which can be either $+\frac{1}{2}$ or $-\frac{1}{2}$, "spin up" or "spin down," for each electron.

The Pauli Exclusion Principle

Each electron has its own unique set of quantum numbers; that is, no two electrons in an atom can have the same four quantum numbers. Two electrons can occupy the same orbital (that is, they can have the same n, ℓ and m_ℓ) but they must have "opposing spins" (that is, different spin quantum numbers; one has $m_s = +\frac{1}{2}$ and the other has $m_s = -\frac{1}{2}$).

You might ask, if an electron has zero probability of being at the node in an orbital, how does it cross the node? Heisenberg's Uncertainty Principle states that we can't observe the electron's motion in any detail. Essentially this question can't be answered--or, in a sense, the question isn't allowed to be asked of a probabilistic model. We could say that the question isn't allowed because the Uncertainty Principle doesn't allow us ever to observe the electron closely enough to see it "move through the node."

The Periodic Table

Problem 7-6:

You are given the following words and numbers. What word(s) should accompany the number 8? What word should accompany number 11?

5: car;   9: ocean liner;   4: "blimp" (or dirigible);   1: glider; 7: commercial passenger jet;   2: bicycle;   6: P.T. boat;   3: canoe; 11: passenger train. (Note: work at this before you look at the solution!)

Solution:

All of these are "modes of transportation"; some in air, some on land, some on water.

a. List the number-word pairs in numerical order.

```
 1.  glider                air
 2.  bicycle               land
 3.  canoe                 water
 4.  "blimp"               air
 5.  car                   land
 6.  P.T. boat             water
 7.  passenger jet         air
 8.  (        )            (    )
 9.  ocean liner           water
10.  (        )            (    )
11.  passenger train       land
```

b. What do you notice? There is a "cycle of threes": Air land water air...

Air		Land		Water	
1.	glider	2.	bicycle	3.	canoe
4.	"blimp"	5.	car	6.	P.T. boat
7.	jet	8.	( )	9.	liner
10.	( )	11.	train		

The columns are "families." As we travel down a column, the vehicle has a larger passenger capacity. Size increases down a column (except for the blimp: There are always exceptions when we look for patterns!).

c. What sort of vehicle is 8?

A land vehicle, carrying more passengers than a car, fewer than a train. A bus?

d. What sort of vehicle is 11?

An air vehicle, carrying more passengers than a commercial jet. Perhaps this vehicle awaits development.

The process of solving this problem is an analogy for the development of the periodic table. Mendeleev discovered patterns in the properties of elements and predicted the properties of "missing" elements corresponding to "holes" in the periodic table.

The quantum mechanical model or electrons in atoms <u>predicts</u> the periodic table, and allows us to extend the periodic table.

Multi-electron Atoms

We have seen orbitals ("wave functions") for one-electron atoms like hydrogen and the helium plus (He^+) ion. This problem can be solved exactly: we only need to consider the attraction between the nucleus and the one electron. The energy of an electron in these atoms depends only on the charge on the nucleus, Z, and the principal quantum number, n, of the orbital.

In a two-electron atom, like the helium atom, we would have to consider the attraction between the nucleus and each of the electrons, and the repulsion between the two electrons. Even a two-electron atom is too complicated to be solved exactly. How will we treat even more complicated atoms?

"Hydrogen-like" Wave Functions

We can use an approximation for multi-electron atoms: if there is an electron between the positively charged nucleus and another electron, the first electron "shields" the second electron from the nucleus, so that the second electron "experiences" a charge smaller than the nuclear charge, the "effective" nuclear charge, Z_{eff}.

As a first approximation, electrons of the same n are, on the average, about the same distance from the nucleus. Electrons of lower n are closer to the nucleus, and would shield the electrons of higher n from the nucleus. Electrons of higher n would not shield electrons of lower n at all.

Then we could use "hydrogen-like" wave functions for each electron in an atom, using Z_{eff} in place of Z in the wave functions.

Replacing Z with Z_{eff} doesn't change the shape of the orbital, but does affect the size of the orbital. Each wave function contains an exponential term:

$$e^{\dfrac{-Zr}{na_0}} \; ;$$

for a given n, as Z, or Z_{eff}, becomes larger, the exponential term decreases more rapidly, and the orbital "shrinks"; as Z, or Z_{eff}, becomes smaller, the orbital becomes larger.

In the hydrogen atoms, $E = -(2.178 \times 10^{-18} \text{ J}) Z^2/n^2$, the energy of an electron in an orbital depends only on n. In a multi-electron atom, electrons in ns orbitals ($\ell = 0$) are slightly lower in energy than electrons in np orbitals ($\ell = 1$), because s electrons penetrate closer to the nucleus than p electrons, and experience a greater "effective nuclear charge".

For each electron in a multi-electron atom:

$$E = -(2.178 \times 10^{-18} \text{ J}) \frac{Z_{eff}^2}{n^2}$$

For two electrons in orbitals of the same n (for example, a 2s orbital and a 2p orbital), the electron "experiencing" the greater "effective nuclear charge" will be lower in energy.

Electron Configurations

We will write descriptions of electrons in an atom, indicating how many electrons occupy each orbital. We will "build up" electron configurations, placing electrons in the lowest energy orbital available, remembering that no two electrons can have the same set of four quantum numbers.

An element with two electrons in a 1s (n = 1, ℓ = 0) orbitals, two electrons in a 2s (n = 2, ℓ = 0) orbital, and six electrons in the three 2p (n = 2, ℓ = 1) orbitals will have the electron configuration: $1s^2 2s^2 2p^6$ (read as "one-s-two, two-s-two, two-p-six"; these superscripts are not powers).

1. Each orbital has "room" for (can be occupied by) two electrons, with "paired spins" (one with m_s = + ½, one with m_s = −½).

2. In multi-electron atom, for a given n, s orbitals are lower in energy than p orbitals, p orbitals are lower in energy than d orbitals, etc.

3. When electrons are placed in a group of empty orbitals of the same energy, the electrons will "half-fill" orbitals, if possible, instead of pairing up with electrons in filled orbitals. This arrangement minimizes repulsions between electrons.

(Hund's rule states that the lowest energy arrangement of single electrons in orbitals of the same energy is the one with the maximum number of unpaired electrons, and with the unpaired electrons having "parallel spins.")

Problem 7-7:

Write the electron configuration for oxygen.

Solution:

a. Oxygen has eight electrons (oxygen's atomic number is 8).

b. Write a blank to indicate an empty orbital, with the label below the blank.

$$\underline{\hspace{1cm}} \quad \underline{\hspace{1cm}} \quad \underline{\hspace{1cm}} \; \underline{\hspace{1cm}} \; \underline{\hspace{1cm}} \quad \underline{\hspace{1cm}} \quad \underline{\hspace{1cm}} \; \underline{\hspace{1cm}} \; \underline{\hspace{1cm}}$$
$$\text{1s} \qquad \text{2s} \qquad\quad \text{2p} \qquad\quad \text{3s} \qquad\quad \text{3p}$$

There are three 2p orbitals of equal energy.

c. Fill the blanks with electrons using our rules:
Use ↑ for an electron with m_s = + ½, ↓ for an electron with m_s = −½.

After five electrons are filled in:

$$\underset{\text{1s}}{\underline{\uparrow\downarrow}} \quad \underset{\text{2s}}{\underline{\uparrow\downarrow}} \quad \underset{\text{2p}}{\underline{\uparrow}\;\underline{\hspace{0.5cm}}\;\underline{\hspace{0.5cm}}} \quad \underset{\text{3s}}{\underline{\hspace{0.5cm}}} \quad \underset{\text{3p}}{\underline{\hspace{0.5cm}}\;\underline{\hspace{0.5cm}}\;\underline{\hspace{0.5cm}}}$$

Where do we place the sixth electron?

Following Hund's rule, the lowest energy configuration would <u>not</u> have two electrons paired in one 2p orbital if other empty 2p orbitals are available.

Therefore, the sixth and seventh electrons would occupy the empty 2p orbitals, with "parallel spins."

After seven electrons:

$$\underset{1s}{\uparrow\downarrow} \qquad \underset{2s}{\uparrow\downarrow} \qquad \underset{2p}{\uparrow \quad \uparrow \quad \uparrow} \qquad \underset{3s}{\underline{}} \qquad \underset{3p}{\underline{} \quad \underline{} \quad \underline{}}$$

In the lowest energy electron configuration the eighth electron must be paired with another electron in a 2p orbital. If we had placed the eighth electron in the empty 3s orbital, we would have a higher energy electron configuration, and "excited state."

The "ground state" (most stable) electron configuration of oxygen is:

$$\underset{1s}{\uparrow\downarrow} \qquad \underset{2s}{\uparrow\downarrow} \qquad \underset{2p}{\uparrow\downarrow \quad \uparrow \quad \uparrow} \qquad \underset{3s}{\underline{}} \qquad \underset{3p}{\underline{} \quad \underline{} \quad \underline{}}$$

or: $1s^2 2s^2 2p^4$, with the distribution of the electrons in the 2p orbitals understood.

The Periodic Table Revisited

The outer electron configuration of elements in the same "family" (the same column in the periodic table) is the same.

All of the elements in column IA have an outer electron configuration ns^1; in IIA, ns^2; in VIIA, $ns^2 2p^5$. The chemistry of the elements involves, for the most part, the electrons in outermost n and p orbitals, the "valence" electrons.

From the periodic table, the chemistry of potassium matches that of sodium and lithium. The ground state electron configuration of K is $1s^2 2s^2 2p^6 3s^2 3p^6 4s^1$, not $1s^2 2s^2 2p^6 3s^2 3p^6 3d^1$. Then, the 4s orbital for potassium must be lower in energy than the 3d orbital.

As a general rule, as we build up the ground state electron configuration for atoms, we will fill (n + 1) s orbitals before nd orbitals. Use the periodic table as your guide.

We can explain trends in ionization energy, electron affinity, and atomic size by looking at the influence of the principal quantum number n, and the effective nuclear charge Z_{eff}, on the energy of the outer electrons, and on the size of the hydrogen-like orbitals.

Look at the periodic table inside the front cover of your text. The elements in columns IA and IIA have an s orbital as the last orbital used in their electron configuration. We could describe these elements as occupying the "s block" of the periodic table. The elements in columns IIIA through VIIA have p orbitals as the last orbitals used in their electron configuration, these elements occupy the "p block." The transition elements, in the B columns, occupy the "d block" and the lanthanides and actinides occupy the "f block."

The periodic table then gives us the following general filling order, as we follow the periodic table, row by row:

1s2s2p3s3p4s3d4p5s4d5p6s4f5d....

Problem 7-8:

Write the electron configuration for I, iodine.

Solution:

I has 53 electrons:

The periodic table indicates the filling order:

We have the electron configuration of argon, after 18 electrons: $1s^2 2s^2 2p^6 3s^2 3p^6$.

The 19th electron would be placed in the 4s orbital (K is in column IA, in the "s" block), the 21st in a 3d orbital (Se is in column IIIB, in the "d" block), the 31st in a 4p orbital (Ga is in column IIIA, in the "p" block), etc.

I: $1s^2 2s^2 2p^6 3s^2 3p^6 4s^2 3d^{10} 4p^6 5s^2 4d^{10} 5p^5$

Postscript

We have spent a lot of time laying the foundations of "quantum chemistry" in the text and in this chapter of the study guide. For now, we will ask you to accept and use a "probabilistic" model for electrons in atoms, even though you may feel uncomfortable with the probabilistic view and the implications of the uncertainty principle. In John Gribbin's book, In Search of Schrödinger's Cat, Niels Bohr is quoted as saying, "Anyone who is not shocked by quantum theory has not understood it."

If you're uncomfortable, you're in good company. You might want to read about the development of quantum theory in other books, including Gribbin's book and Taking the Quantum Leap by Fred Alan Wolf, among others.

CHAPTER SEVEN: SELF-TEST

7-1.   a.   What is the frequency of light with a wavelength of 485 nm?

b.   What is the energy of a photon of light with this wavelength?

7-2.   What is the de Broglie wavelength of a 0.250 kg ball travelling at 17.5 meters per second?

7-3.   The energy of an electron with principal quantum number n in a hydrogen atom is:

$$E_n = -2.178 \times 10^{-18} \text{ J} \times \frac{1}{n^2}$$

i)   Is energy emitted or absorbed when an electron "jumps" from an energy level with n = 2 to an energy level with n = 5?

ii)   What is the wavelength of the photon associated with an electron "jumping" from n = 2 to n = 5?

7-4.   Sketch the shapes of

a.   A 3s orbital

b.   A 2p orbital

c.   A $3d_{z^2}$ orbital

d.   A $3d_{xy}$ orbital

7-5.   How many electrons in an atom can have

a.   n = 2

b.   n = 3 and ℓ = 2

c.   n = 4 and ℓ = 3 and m_s = +1/2

d.   n = 3 and ℓ = 1 and m_ℓ = -1

7-6.   Write the electron configurations for

a.   Ca

b.   Zn

c.   S

d.   Sn

7-7.   Chromium has an electron configuration with 6 unpaired electrons. Write an electron configuration for Cr.

7-8. How many unpaired electrons are there in:

    a. O

    b. Ca

    c. Br

7-9. In each pair, choose the atom with the largest radius.

    a. C F

    b. C Si

    c. F S

7-10. In each pair, choose the atom which loses an electron most easily.

    a. Mg Na

    b. F Cl

    c. Mg K

7-1. a. $\nu = c/\lambda = 6.18 \times 10^{14}$ s^{-1}

b. $E = h\nu = 4.09 \times 10^{-19}$ J

7-2. $\nu = h/mv = 1.51 \times 10^{-34}$ m

7-3. $E_2 = -5.445 \times 10^{-19}$ J

$E_5 = -8.712 \times 10^{-20}$ J

i) Energy is absorbed.

ii) $\Delta E = 4.574 \times 10^{-19}$ J

$\lambda = 4.343 \times 10^{-7}$ m $= 434.3$ nm

7-4. See the drawings in the text.

7-5. a. 8 electrons: 2 s electrons and 6 p electrons

b. 10 3d electrons

c. m_ℓ can be $-3, -2, -1, 0, +1, +2, +3$; m_s must be $+1/2$: 7 electrons

d. $n = 3$, $\ell = 1$, $m_\ell = -1$: m_s can be $+1/2$ or $-1/2$: 2 electrons

7-6. a. Ca: $1s^2 2s^2 2p^6 3s^2 3p^6 4s^2$

b. Zn: $1s^2 2s^2 2p^6 3s^2 3p^6 4s^2 3d^{10}$

c. S: $1s^2 2s^2 2p^6 3s^2 3p^4$

d. Sn: $1s^2 2s^2 2p^6 3s^2 3p^6 4s^2 3d^{10} 4p^6 5s^2 4d^{10} 5p^2$

7-7. Cr: $1s^2 2s^2 2p^6 3s^2 3p^6 4s^1 3d^5$ (Not $[Ar]4s^2 3d^4$)

7-8. a. O: $1s^2 2s^2 2p^4$ $\underline{\uparrow\downarrow}$ $\underline{\uparrow}$ $\underline{\uparrow}$ 2 unpaired electrons
$\quad$ 2p

b. Ca: $1s^2 2s^2 2p^6 3s^2 3p^6 4s^2$ $\underline{\uparrow\downarrow}$ 0 unpaired electrons
$\quad$ 4s

c. Br: $1s^2 2s^2 2p^6 3s^2 3p^6 4s^2 3d^{10} 4p^5$ $\underline{\uparrow\downarrow}$ $\underline{\uparrow\downarrow}$ $\underline{\uparrow}$ 1 unpaired electron
$\quad$ 4p

7-9. a. C > F

b. Si > C

c. S > O > F

7-10. a. Na > Mg

b. Cl > F

c. K > Na > Mg

CHAPTER EIGHT: CHEMICAL BONDING

Spontaneity

Why do compounds form? What holds atoms together in compounds? Behind these questions is the even more fundamental question: why is a process spontaneous? We'll look at this question again in Chapter 16. For now, we'll use two qualitative rules: first, there is a natural tendency for a system to achieve a state of lower potential energy. For example, a ball rolls down hill, and comes to rest: the ball's potential energy decreases, and friction converts the kinetic energy of the ball into microscopic kinetic energy of the atoms and molecules in the ball and the surface it rolls on. We never see a ball at rest convert microscopic kinetic energy into potential energy; that is, the ball doesn't start at rest, roll up the hill, leaving the surface cooler.

There is a second natural tendency, for systems to achieve a state of "more disorder," or a more probable arrangement. For example, gases expand into an evacuated container. Qualitatively, every spontaneous process must either be accompanied by a decrease in potential energy or lead to a more probable arrangement of particles.

What does this have to do with "why compounds form"?

- When sodium metal and chlorine gas react to form sodium chloride, the crystalline product is a very ordered array of positive sodium ions and negative chloride ions, more ordered than the reactants.

- If every spontaneous reaction must either be exothermic or lead to a more probable arrangement of atoms in the products, then the crystalline product won't form unless the reaction is exothermic.

We can break the process of forming NaCl down into simple steps to see why the process is exothermic. We find that ionic compounds form because of the strong attraction between positive and negative ions in the crystal.

When two hydrogen atoms form a hydrogen molecule, the product is a "less probable" arrangement of atoms. Therefore, the product won't form unless the process is exothermic, that is, the two atoms in the molecule have a lower potential energy than the separated atoms. We find that these atoms are held together by the attractive forces between negatively charged electrons and positively charged nuclei, balanced by the repulsions between the electrons and repulsions between the nuclei.

Electronegativity and Polarity

When a pair of electrons is shared between two like atoms, the pair is shared equally. Electronegativity is a measure of the attraction of an atom for electrons in a bond. If a pair of electrons is shared between two atoms of different electronegativity, the bonding electrons will be shared unequally. The more electronegative atom will have a slightly negative charge ($\delta-$), the less electronegative atom will have a slight positive charge ($\delta+$), and the bond will be polar.

Electronegativity increases to the right across a row in the periodic table, and decreases down a column in the periodic table. Fluorine is the most electronegative atom, with EN = 4.0.

The greater the difference in electronegativities between two atoms in a bond, the more polar the bond.

Problem 8-1:

Without looking at Figure 8.3 in the text, which of the bonds in each of these sets would be more polar? Remember the trends.

    i.   ICl   BrCl

    ii. CO   CN

    iii. NO   PO

Solution:

    i.   Down a column, electronegativity decreases:

         Therefore, $EN_{Cl} > EN_{Br} > EN_I$

         The difference in electronegativity between I and Cl is greater than the difference between Br and Cl. ICl is the more polar bond.

    ii. Across a row, electronegativity increases:

         $$EN_O > EN_N > EN_C$$

         CO is the more polar bond.

    iii. Using the row trend: $EN_O > EN_N$

         Using the column trend: $EN_N > EN_P$

         Therefore, $EN_O > EN_N > EN_P$

         And PO is the more polar bond.

Polar Molecules

To determine whether a molecule is polar we must know whether the bonds in the molecule are polar and we must know the geometry of the molecule. Later in this chapter we will use a model to predict geometries of molecules. For now, we will be given the geometry of the molecule.

We can use two different methods to determine whether a molecule is polar. The first method uses "vector addition." Begin by drawing arrows along polar bonds, ↔, with the head of the arrow pointing toward the negative end of the bond dipole. Then "add" these arrows (vectors), to find the net result. For example, for the water molecule: with the net:

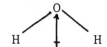

In the second method, begin by writing $\delta-$ at the negative end and $\delta+$ at the positive end of every polar bond. For example, for water

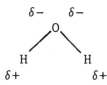

Then, find the "center-of-positive-charges" and mark it with an x; find the "center-of-negative-charges" and mark it with a dot (•)

If the center-of-positive-charges is at a different place than the center-of-negative-charges, the molecule is polar. The dot (•) marks the negative end of the molecule, and the x marks the positive end.

Each method gives the same answer for the polarity of a molecule.

Problem 8-2:

   CH_2F_2 is a tetrahedral molecule. Is CH_2F_2 polar?

Solution:

   a. Does CH_2F_2 have polar bonds? Look at the electronegativities:

   $$EN_C = 2.5; \quad EN_F = 4.0; \quad EN_H = 2.1$$

   The C–F bonds are clearly more polar than the C–H bonds.

   b. Draw the geometry of the molecule. The tetrahedral molecule can be drawn, where ——— represents a bond in the plane of the paper, --- represents a bond projecting behind the plane of the paper, and ◄── represents a bond projecting forward.

117

Locate $\delta+$ and $\delta-$ at the ends of each bond. Find the approximate centers of positive charge (x) and negative charge (•).

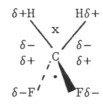

This molecule is polar.

You may find it easier to look at a 3-dimensional model of the molecule.

Ions

You should be able to predict what ion an element will form, using the generalization in the text that metal atoms will lose electrons to form a cation with a noble gas electron configuration and nonmetal atoms will gain electrons with a noble gas electron configuration.

Problem 8-3:

What are the electron configurations for the ions formed from each of the following atoms:

    i. Mg   ii. S   iii. Br   iv. K

Solution:

Write the electron configuration for each atom, and then for the ion.

    i. Mg: $1s^2 2s^2 2p^6 3s^2$

        Magnesium loses two electrons to form Mg^{2+}, with same electron configuration as neon: $1s^2 2s^2 2p^6$.

   ii. S: $1s^2 2s^2 2p^6 3s^2 3p^4$

        Sulfur gains two electrons to form S^{2-}, with the same electron configuration as argon: $1s^2 2s^2 2p^6 3s^2 3p^6$.

  iii. Br: $1s^2 2s^2 2p^6 3s^2 3p^6 4s^2 3d^{10} 4p^5$

        Bromine gains one electron to form Br^-: $1s^2 2s^2 2p^6 3s^2 3p^6 4s^2 3d^{10} 4p^6$.

   iv. K: $1s^2 2s^2 2p^6 3s^2 3p^6$

        K^+: $1s^2 2s^2 2p^6 3s^2 3p^6$

K^+ and S^{2-} are isoelectronic: that is, they have the same electron configuration. Which is smaller? If they have the same electron configuration, the size of the ion depends only on the charge on the nucleus. The nucleus of the potassium atom has a +19 charge; the nucleus of the sulfur atom has a +16 charge. Therefore, the potassium cation is smaller than the sulfide anion.

Ionic size decreases down a column (group) in the periodic table.

Ionic Compounds

We can break the formation of a crystalline ionic compound from its elements into a series of steps. The enthalpy change for the overall process is the sum of the enthalpy changes for each of the steps. Because our product is "highly ordered," we know that the reaction will only take place if the overall process is exothermic.

Look at Figure 8.9 in the text: Step 1, 2, and 3 each "cost" energy. Some of the "energy cost" is paid by step 4, adding an electron to the fluorine atom. However, the process of forming gaseous ions from solid lithium and fluorine gas is endothermic, that is, the process requires energy. Step 5, forming solid LiF from the gaseous ions pays all the "energy costs" of forming the ions and produces 617 kJ/mole to be released to the surroundings.

The "driving force," then, for forming ionic compounds is the lattice energy, the energy released when the charged ions combine to form the solid.

Bonds

Our idea of "bonds" is really a model, rather like the ball-and-stick model sets you may use in the laboratory. Every model is a simplified view of reality. In every model, there are some aspects of the model that are "true," that is, match "reality," and others that don't. For example, when you build a ball-and-stick model of methane or of water, you use hard colored spheres for each atom, with rods to represent pairs of electrons. Atoms are not hard spheres, and have no color. Electrons aren't solid rods. The "true" aspects of the model are only the geometry of the molecule and the approximate angles between bonds in the molecule. Ball-and-stick models are useful to help us visualize three-dimensional shapes of molecules. However, don't be misled into thinking that molecules actually look like their ball-and-stick models. Remember that models are a simplification. Each model has its strengths and weaknesses.

One model of chemical bonds assumes that bonds in molecules are independent of each other, that a hydrogen-carbon bond in one molecule is very much like a hydrogen-carbon bond in another. Then, we can calculate the energy associated with a particular bond and use this energy in any molecule.

If we use this model of bonds in molecules, we can look at any chemical reaction in a step-wise fashion, breaking bonds in the reactants and forming new bonds in the products. This isn't the mechanism of the reaction. However, enthalpy is a state function, and the change in enthalpy for a reaction cannot depend on the pathway we choose. We can use the changes in enthalpy for these steps to calculate the overall charge in enthalpy.

Problem 8-4:

Use bond energies to calculate the change in enthalpy expected when methane, CH_4, burns in oxygen to form carbon dioxide, CO_2, and water.

Solution:

We will use the formulas given for these molecules. Later in this chapter we will learn to develop these formulas:

$$
\begin{matrix}
& H & \\
& | & \\
H - & C & - H \quad + \quad 2\ O \!=\! O \quad \rightarrow \quad O \!=\! C \!=\! O \quad + \quad 2\ H - O - H \\
& | & \\
& H &
\end{matrix}
$$

Energy is required to break the bonds in the reactants, and energy is released when bonds form in the products.

a. Calculate the energy to break the bonds in the reactants use the bond energies in Table 8.4:

We break 4 C – H bonds: 4(413 kJ) = 1652 kJ
   and 2 O O bonds: 2(495 kJ) = 990 kJ
         with a net "energy cost" +2642 kJ

b. Calculate the energy released when bonds are formed in the products:

We form: 2 C O bonds   2(799 kJ) = 1598 kJ
       4 H – O bonds   4(467 kJ) = 1868
       with a net "energy released" +3466 kJ

c. Calculate the total change in enthalpy:

ΔH = energy to break old bonds – energy released when new bond form.

   = 2642 kJ – 3466 kJ = – 824 kJ

That is 824 kJ are released when one mole of methane "burns" in oxygen.

The Localized Electron Model

This model assumes that we can think of a molecule as held together by pairs of shared electrons, a pair of shared electrons acts as a bond between two atoms. We will develop this model in steps.

120

Lewis Structures

To write Lewis structures we use a shorthand notation for the atom. We will use the symbol for the atom to represent the nucleus of the atom and its inner core of electrons, and dots to represent the atom's valence electrons. (This model, with electrons represented as fixed dots, seems inconsistent with our quantum mechanical view of electrons and atoms. Indeed, it is.)

Problem 8-5:

Write Lewis structures for each of the following atoms and ions:

    i. H   ii. C   iii. O   iv. Cl^- v. Na^+ vi. Se

Solution:

    i.   Hydrogen has the electron configuration $1s^1$. There is one valence electron, and the Lewis structure is H·

   ii.   Carbon has the electron configuration: $1s^2 2s^2 2p^2$, with four valence electrons ($2s^2 2p^2$), the Lewis structure is C

  iii.   Oxygen has the electron configuration: $1s^2 2s^2 2p^4$, with six valence electrons: O

   iv.   Chlorine has the electron configuration: $1s^2 2s^2 2p^6 3s^2 3p^5$
       Chloride, Cl^-, has the electron configuration: $1s^2 2s^2 2p^6 3s^2 3p^6$ with eight valence electrons. Cl $^-$

    v.   Sodium has the electron configuration: $1s^2 2s^2 2p^6 3s^1$
       Sodium ion, Na^+ has the electron configuration: $1s^2 2s^2 2p^6$
       Na^+ might be thought of as having eight valence electrons. However, usually we represent metal ions with noble gas electron configurations as having no valence electrons: Na^+

   vi.   Selenium has the electron configuration: $1s^2 2s^2 2p^6 3s^2 3p^6 4s^2 3d^{10} 4p^4$ Only the outermost s and p electrons are counted as valence electrons. Selenium has six valence electrons: Se

In the Lewis model of bonding, atoms share pairs of electrons to form bonds. Atoms will share electrons to achieve a "noble gas electron configuration" of valence electrons. That is, hydrogen will share electrons to achieve a "complete duet" of two valence electrons; oxygen will share electrons to achieve a "complete octet" of eight valence electrons. Each atom in a bond counts all the shared electrons in the bond in its "valence shell." The justification for this simple model is its ability to predict or explain stable molecules.

We will begin to write Lewis structures for molecules where we are given a "bonding skeleton." For example, to write a Lewis structure for NH_3, we would be told that the three hydrogen atoms are each bound to the nitrogen atom.

Follow these steps to write a Lewis structure for NH_3

1. Write the bonding skeleton

        H   N   H

            H

2. Count the number of valence electrons available:

    Each hydrogen has one: H   3(1) = 3

    The nitrogen has five: N   = 5

            total valence electrons   8 or 4 pairs

3. Use pairs of electrons for each bond in the bonding skeleton:
    Represent a pair of electrons with a pair of dots (••) or a dash (–)

        H – N – H
            |
            H

    We've used three pairs of electrons, and have one pair remaining.

4. Use the remaining electron pairs to complete the valence shells of
    any atoms with incomplete valence shells ("octets" for all but
    hydrogen)

            ••
        H – N – H
            |
            H

5. Check that every atom has a complete valence shell.

    NH_3 has three N – H bonds, and a non-bonding ("lone") pair of elec-
    trons on the nitrogen.

Now let us write a Lewis structure for ethylene, C_2H_4.

1. We are given the bonding skeleton:

        H   C   C   H

            H   H

2. There are 12 valence electrons available

        2 C = 8

        4 H = 4
                 12 or 6 pairs

3. We use 5 electron pairs to complete the bonding skeleton: one pair remains.

```
H - C - C - H
    |   |
    H   H
```

4. We can use the remaining pair of electrons to complete the octet for one carbon. The other carbon has an "incomplete octet." We need a new rule for the occasions when there are not enough electron pairs to "complete" all the "incomplete" octets.

```
    ..
H - C - C - H
    |   |
    H   H
```

4a. If there are still incomplete octets, share the lone pair of electrons on one atom with another atom to form a multiple bond.

```
    .. ↷                              
H - C - C - H      to give     H - C ≡ C - H
    |   |                          |   |
    H   H                          H   H
```

with a double bond (two shared pairs of electrons) between the two carbon atoms. Double bonds are shorter and stronger than single bonds.

Problem 8-6:

Write Lewis structures for the following molecules and ions.

a. H_2CO

1. The bonding skeleton

```
H   C   O

H
```

2. Valence electrons:

$$2 (H) = 2$$

$$C = 4$$

$$O = \underline{6}$$

$$12 \text{ valence electrons}$$

3. Use three bonding pairs

```
H - C - O
    |
    H
```

4. Three electron pairs remain. Both carbon and oxygen have "incomplete octets." As a general rule, fill the octets of more
electronegative atoms, in this case oxygen, first

$$H - \overset{\displaystyle |}{\underset{\displaystyle |}{C}} - \overset{\displaystyle ..}{\underset{\displaystyle ..}{O}}:$$
$$\quad\quad H$$

4a. Carbon still has an incomplete octet: Share a lone pair from
oxygen

$$H - \overset{\displaystyle |}{\underset{\displaystyle |}{C}} - \overset{\displaystyle ..}{\underset{\displaystyle ..}{O}}: \quad\quad \text{to give} \quad\quad H - \overset{\displaystyle |}{\underset{\displaystyle |}{C}} = \overset{\displaystyle ..}{O}:$$
$$\quad\quad H \quad\quad\quad\quad\quad\quad\quad\quad\quad\quad H$$

5. Check that each atom has a complete valence shell.

b. SO_4^{2-}, sulfate ion

1. The bonding skeleton:

$$O$$
$$O \quad S \quad O$$
$$O$$

2. The valence electrons

$$1\ S = 6$$

$$4\ (O) = 24$$

Plus 2 two electrons for the −2 charge $\quad\underline{2}$
$$\quad\quad\quad\quad\quad\quad\quad\quad\quad\quad\quad 32 \text{ electrons, or}$$
$$\quad\quad\quad\quad\quad\quad\quad\quad\quad\quad\quad 16 \text{ pairs}$$

3. Use 4 pairs for bonding:

$$\overset{\displaystyle O}{\underset{\displaystyle |}{}}$$
$$O - S - O$$
$$\underset{\displaystyle O}{\overset{\displaystyle |}{}}$$

12 pairs remain:

4. Use the remaining pairs to complete octets. Sulfur's octet is
complete:

$$\overset{\displaystyle ..}{\underset{\displaystyle |}{O}}: \quad\quad 2-$$
$$:\overset{\displaystyle ..}{\underset{\displaystyle ..}{O}} - S - \overset{\displaystyle ..}{\underset{\displaystyle ..}{O}}:$$
$$\underset{\displaystyle :\overset{\displaystyle }{\underset{\displaystyle ..}{O}}:}{\overset{\displaystyle |}{}}$$

5. Check that each atom has a complete valence shell.

c. $CO_3{}^{2-}$, carbonate ion.

1. Bonding skeleton

$$O \quad C \quad O$$
$$O$$

2. Valence electrons

$$C = 4$$
$$3(O) = 18$$
$$-2 \text{ charge} = \underline{2}$$
$$24 \text{ electrons; 12 pairs}$$

3. Use 3 pairs to form the bonding skeleton: 9 pairs remain

$$O - C - O$$
$$|$$
$$O$$

4. Complete the octets of the more electronegative atoms first:

$$:\overset{..}{O} - C - \overset{..}{\underset{..}{O}}: \quad {}^{2-}$$
$$\underset{..}{|}$$
$$:\overset{..}{\underset{..}{O}}:$$

4a. Carbon's octet is still incomplete: Share a lone pair from an oxygen.

$$:\overset{\frown}{O} - C - \overset{..}{\underset{..}{O}}: \quad {}^{2-} \qquad \text{to give} \qquad :O = C - \overset{..}{\underset{..}{O}}: \quad {}^{2-}$$
$$\underset{..}{|} \qquad\qquad\qquad\qquad\qquad\qquad |$$
$$:\overset{..}{\underset{..}{O}}: \qquad\qquad\qquad\qquad\qquad\qquad :\underset{..}{O}:$$

5. Check that each atom has a complete valence shell.

This structure for the carbonate ion predicts that $CO_3{}^{2-}$ will have two long carbon-oxygen single bonds, and one short carbon-oxygen double bond. However, the carbon-oxygen bonds in $CO_3{}^{2-}$ are each the same length and each is shorter than a single bond and longer than a double bond.

Let's look again at possible Lewis structures for $CO_3{}^{2-}$. We could have written three equivalent electron arrangements for the same arrangement of atoms:

$$:O = C - \overset{..}{\underset{..}{O}}: \quad {}^{2-} \qquad :\overset{..}{O} - C - \overset{..}{\underset{..}{O}}: \quad {}^{2-} \qquad :\overset{..}{O} - C = O: \quad {}^{2-}$$
$$| \qquad\qquad\qquad\qquad || \qquad\qquad\qquad\qquad |$$
$$:\underset{..}{O}: \qquad\qquad\qquad\qquad :\underset{..}{O}: \qquad\qquad\qquad\qquad :\underset{..}{O}:$$

Whenever we can write more than one electron arrangement for the same arrangement of atoms, the actual molecule appears to be an average. We call these arrangements "resonance structures" and connect them with double-headed arrows ($\leftrightarrow$).

We'll add step 5b to our procedure for writing Lewis structures.

5b. Look for other possible equivalent arrangements of electrons and write resonance structures.

We have to "invoke" resonance structures to patch up our localized electron model. We could envision the pair of electrons forming the localized double bond in one $CO_3{}^{2-}$ resonance structure as "delocalized" over all three resonance forms.

Exceptions

For some compounds, for example, BeH_2 and BF_3, we write Lewis structures with a small central atom with an "incomplete octet."

In other compounds, we are forced to write a Lewis structure with the central atom having an "expanded octet," that is, a valence shell with more than eight electrons. We only find expanded octets for non-metal atoms beyond row 2 in the periodic table, and then, only when the central atom is bonded to small electronegative atoms such as fluorine, oxygen, and chlorine.

Valence Shell Electron Pair Repulsion

The VSEPR model lets us predict shapes of molecules from their Lewis structures. Look at the "sets" of electrons around an atom. A pair of electrons in a single bond is a "set"; both pairs of electrons in a double bond act as one "set." Each set must remain together to form a bond. But sets will tend to be as far apart from each other as possible.

Example	Sets of Electrons	Bond Angles	Arrangement of Electron Sets	Shorthand Drawing
Cl – Be – Cl	2	180°	linear	
F$\diagdown$B$\diagup$F $\vert$ F	3	120°	trigonal (or triangular) planar	

	Example	Sets of Electrons	Bond Angles	Arrangement of Electron Sets	Shorthand Drawing

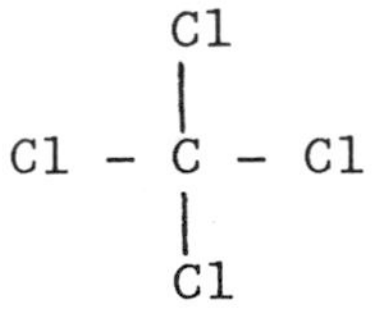

```
        Cl
        |
  Cl  - C - Cl       4            109°         tetrahedral
        |
        Cl
```

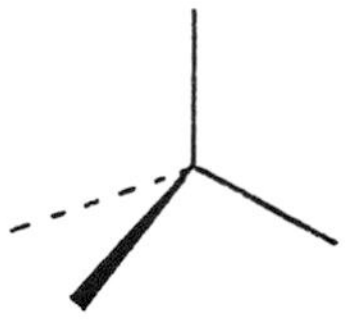

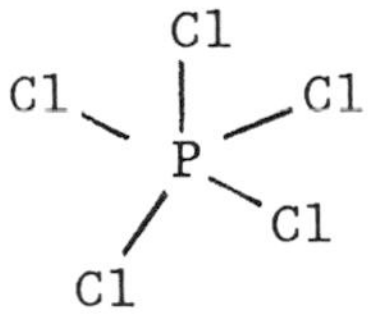

```
        Cl
  Cl\   |  /Cl                     90°
      \ P /           5             &          trigonal
       / \Cl                       120°        bipyramid
  Cl /
```

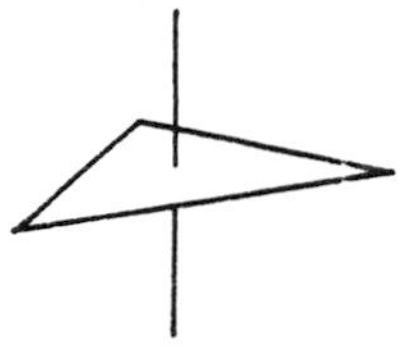

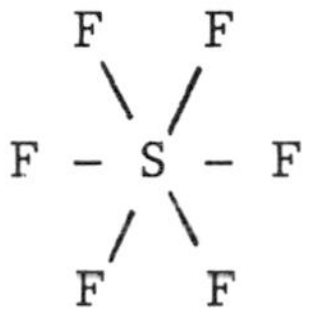

```
   F   F
    \ /
  F - S - F           6             90°         octahedral
    / \
   F   F
```

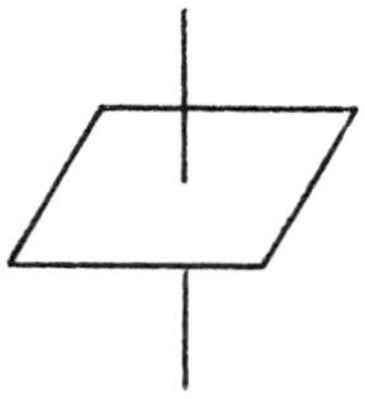

To predict the geometry of a molecule:

1. Draw the Lewis structure

2. Arrange the sets of bonding and non-bonding electrons around each
    atom to minimize repulsion between sets.

[2a. Remember that non-bonding ("lone") pairs of electrons tend to occupy
     more space than bonding pairs. Lone pairs tend to force bonding
     pairs closer together. Two lone pairs in octahedral and trigonal
     bipyramidal arrangements of electrons tend to occupy positions as far
     apart from each other as possible.]

3. Describe the geometry of the molecule from the positions of the
    atoms, not the sets of electrons.

	Example	Arrangement of Electrons	Bond Angles	Molecular Geometry
	H - O - H	Tetrahedral	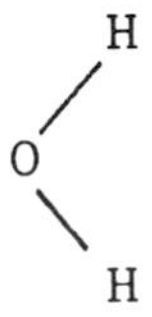 O with H's, <109°	Bent

127

| | Arrangement | Bond | Molecular |
Example	of Electrons	Angles	Geometry

H – N̈ – H   Tetrahedral   <109°   Trigonal-based pyramid (pyramidal)

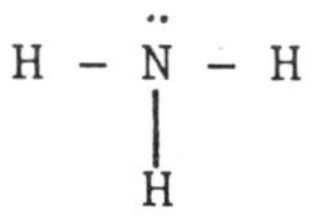

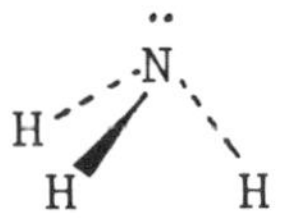

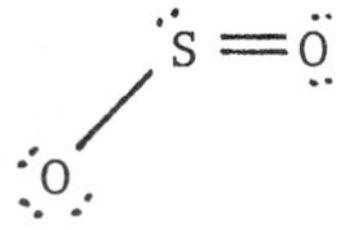

$\overset{|}{H}$

S̈ ＝ Ö   Trigonal planar   :S   120°   Bent

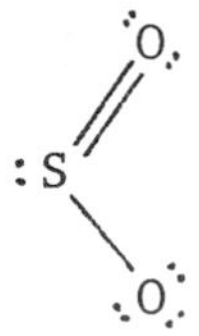

4. For complicated molecules, describe the bond angles

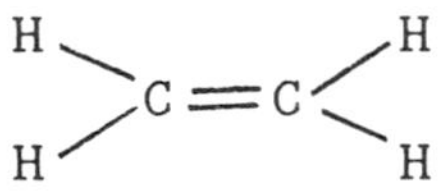

H, H   H
 C ＝ C   trigonal planar   C – H bond angle is 120° 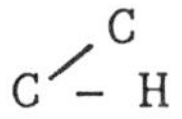
H, H   about each carbon

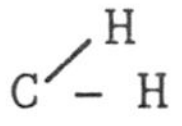

C – H bond angle is 120°

This molecule is planar, although there is nothing about its Lewis structure that requires it to be planar. We'll discover why this molecule is planar in Chapter 9.

Problem 8-7:

Describe the geometry of acetic acid.

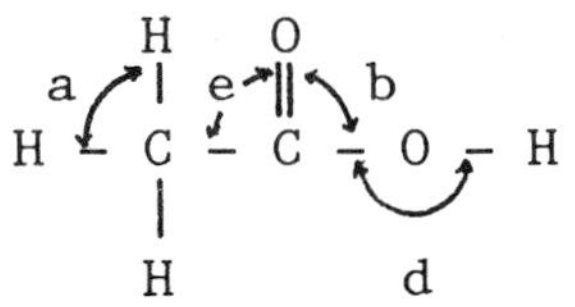

H   O
a ⟋| e ‖ ↘ b
H ⟵ C – C ↓ O – H
|
H   d

Solution:

We can describe the geometry of this complex molecule by describing all its bond angles:

a: C – H   There are four sets of electrons around the left hand carbon: The electrons are arranged tetrahedrally and the bond angle is ≈ 109°.

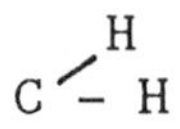

e:
C—O (with O above, structure: C—C with O attached to left C)

There are three sets of electrons around the right hand carbon: The electrons are in a trigonal planar arrangement and the Se bond angles are 120°.

and

b:
C—O (with O above left C)

d:
O—H (with C above left O)

There are four sets of electrons around the right hand oxygen: The electrons are in a tetrahedral arrangement. The two lone pairs on the oxygen atom will tend to force the bond angle to be less than 109°.

Now that you can predict molecular geometries, you can predict the polarity of molecules.

CHAPTER EIGHT: SELF-TEST

8-1. Which of the bonds in each pair is more polar?

    a. NO   NF

    b. FBr   FI

    c. SO   SF

8-2. Write the electron configuration for the following ions:

    a. Al^{3+}

    b. Br^-

    c. Fe^{2+}

    d. S^{2-}

8-3. Which ion in each set is the largest? Which is the smallest?

    a. O^{2-}, F^-, Na^+

    b. Mg^{2+}, Na^+, K^+

8-4. Use bond energies to calculate ΔH for the reaction:

$$2 H_2(g) + O_2(g) \rightarrow 2 H_2O(g)$$

Bond	Bond Energies
H – H	436 kJ/mole
O O	495
H – O	467

8-5. Write Lewis structures for the following atoms, molecules and ions:

    a. Br^-

    b. Mg^{2+}

    c. NF_3 (Each fluorine is bound to the central nitrogen atom.)

    d. ClO_4^-

    e. ClO_2^-

    f. CO_3^{2-}

    g. HCN

8-6. Give the shapes for the species above:

 c. NF_3

 d. ClO_4^-

 e. ClO_2^-

 f. CO_3^{2-}

 g. HCN

8-7. Given these Lewis structures, decide whether each molecule is polar or non-polar.

a. 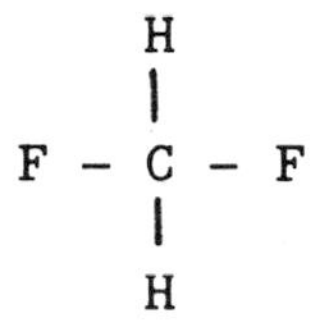
$$\begin{array}{c} H \\ | \\ F - C - F \\ | \\ H \end{array}$$

b. 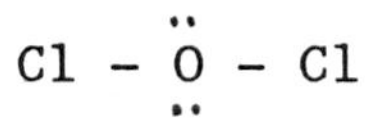 $Cl - \overset{\cdot\cdot}{\underset{\cdot\cdot}{O}} - Cl$

c.
$$\begin{array}{c} F - B - F \\ | \\ F \end{array}$$

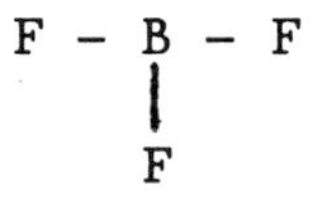

8-1. a. NF

b. FI

c. SF

8-2. Al^{3+}: $1s^2 2s^2 2p^6$

Br^-: $1s^2 2s^2 2p^6 3s^2 3p^6 4s^2 3d^{10} 4p^6$

Fe^{2+}: $1s^2 2s^2 2p^6 3s^2 3p^6 3d^6$

S^{2-}: $1s^2 2s^2 2p^6 3s^2 3p^6$

8-3. a. These ions are isoelectronic, each with the electron configuration $1s^2 2s^2 2p^6$: Na^+ is smallest, with the largest nuclear charge; O^{2-} is the largest.

b. Mg^{2+} and Na^+ are isoelectronic: Mg^{2+} is the smaller. K^+ is larger than Na^+: K^+ is the largest ion.

8-4. $$2 H - H + O = O \rightarrow 2 H - O - H$$

ΔH = energy to break bonds in reactants – energy released when bonds are made in products.

= 2(436 kJ) + 495 kJ – 4(467 kJ)

= –501 kJ

8-5. a. Br^- 8 valence electrons

b. Mg^{2+} 0 valence electrons

c. 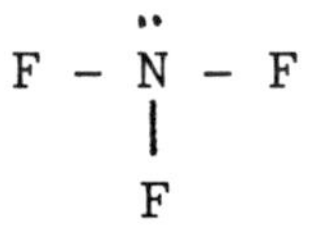 F – N̈ – F
       |
       F

d. :O:
       |
   :Ö – Cl – Ö:
       |
      :O:

e. :Ö – Cl̈ – Ö: ⁻

f. :Ö – C ＝ Ö: ⇌ :Ö – C – Ö: ⇌ :Ö ＝ C – Ö: (each with :Ö: at top, charge 2–)

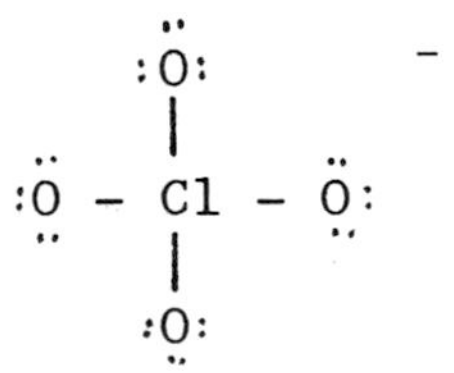

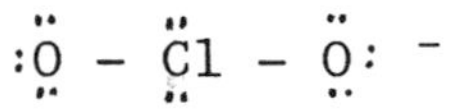

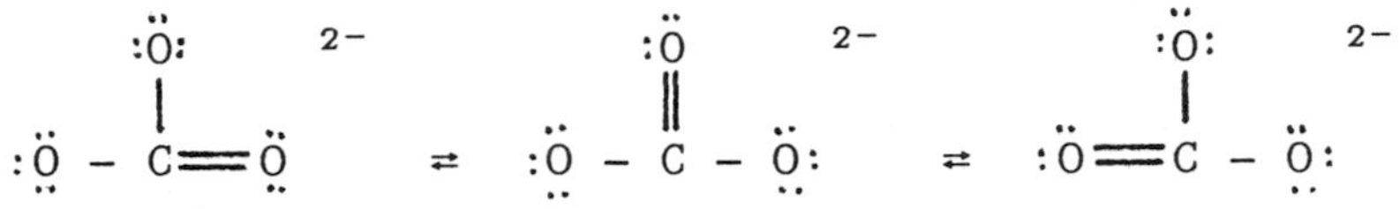

g. H – C ≡ N

8-6. c. pyramidal

        d. tetrahedral

        e. bent

        f. triangular planar

        g. linear

8-7. a. polar: the molecule is tetrahedral.

        b. polar: the molecule is bent.

        c. non-polar: the molecule is triangular planar.

CHAPTER NINE: COVALENT BONDING

The Lewis model of bonding seems far removed from the quantum theory we looked at in Chapter 7. Let's look at another approach to the localized electron model. This model considers bonds as formed by the overlap of a half-filled atomic orbital on one atom with a half-filled orbital on another atom (or, less frequently, by the overlap of a filled atomic orbital on one atom with an empty atomic orbital on another atom).

Problem 9-1:

Describe the orbitals which overlap to form bonds in

   i) H_2
  ii) F_2
 iii) HF
  iv) O_2

Solution:

Write the electron configuration for each atom in the bond.

  i) H_2: H $\underset{1s}{\uparrow}$

        H $\underset{1s}{\uparrow}$

Then overlap the two half-filled 1s orbitals to form the bond.

This bond is cylindrically symmetric. In cross section it resembles an s orbital. A cylindrically symmetric bond is called a "sigma" (σ) orbital.

  ii) F_2: F $\underset{1s}{\uparrow\downarrow}$ $\underset{2s}{\uparrow\downarrow}$ $\underset{}{}$ $\underset{2p}{\uparrow\downarrow}$ $\underset{}{\uparrow\downarrow}$ $\underset{}{\uparrow}$

        F $\underset{1s}{\uparrow\downarrow}$ $\underset{2s}{\uparrow\downarrow}$ $\underset{}{}$ $\underset{2p}{\uparrow\downarrow}$ $\underset{}{\uparrow\downarrow}$ $\underset{}{\uparrow}$

Overlap the two half-filled p orbitals end-to-end.

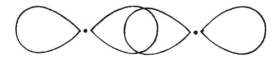

This bond is also cylindrically symmetric, a σ bond.

iii) HF: H   ↑
               ‾‾
               1s

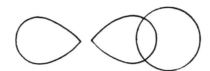

F   ↑↓   ↑↓   ↑↓   ↑↓   ↑
     ‾‾   ‾‾   ‾‾   ‾‾   ‾‾
     1s   2s   2p

Overlap the half-filled 1s orbital on the hydrogen atom with the half-filled 2p orbital on fluorine to form a σ bond.

iv) O₂: O   ↑↓   ↑↓   ↑↓   ↑   ↑
                ‾‾   ‾‾   ‾‾   ‾‾   ‾‾
                1s   2s   2p

           O   ↑↓   ↑↓   ↑↓   ↑   ↑
                ‾‾   ‾‾   ‾‾   ‾‾   ‾‾
                1s   2s   2p

We can make two bonds between the oxygen atoms: first, a σ bond from, "end-to-end," overlap of one half-filled atomic orbital on each oxygen.

The other half-filled p orbitals are perpendicular to those forming the σ bond. These p orbitals can overlap side-to-side.

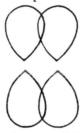

In cross-section, this bond resembles a p orbital. There is a node along the line connecting the two nuclei. This is a "pi" (π) bond.

Let's look at more complicated molecules. How would we describe the bonding in water?

First, write the electron-configurations for each atom:

H   ↑
     ‾‾
     1s

H   ↑
     ‾‾
     1s

O   ↑↓   ↑↓   ↑↓   ↑   ↑
     ‾‾   ‾‾   ‾‾   ‾‾   ‾‾
     1s   2s   2p

We could overlap a half-filled p orbital on the oxygen atom with the half-filled 1s orbital on one hydrogen atom and overlap the other half-filled p orbital on the oxygen atom with the half-filled 1s orbital on the other hydrogen atom. Each bond would be a σ bond:

However, this model predicts a bond angle of 90°, not the actual 104°.

Let's look at another molecule, CH_4.

The electron configurations are:

each H $\quad$ $\underline{\uparrow}$
$\qquad\qquad$ 1s

C $\quad$ $\underline{\uparrow\downarrow}$ $\underline{\uparrow\downarrow}$ $\quad$ $\underline{\uparrow}$ $\underline{\uparrow}$ $\underline{}$
$\qquad$ 1s $\quad$ 2s $\qquad\quad$ 2p

We might expect to form CH_2 by overlapping half-filled orbitals. To form four bonds, we need four half-filled orbitals.

We could create four half-filled orbitals by "exciting" an electron from the 2s orbital on the carbon atom to the 2p orbital. This would cost energy, but the cost might be repaid by the extra energy released when two more bonds form.

C $\quad$ $\underline{\uparrow\downarrow}$ $\underline{\uparrow}$ $\quad$ $\underline{\uparrow}$ $\underline{\uparrow}$ $\underline{\uparrow}$
$\qquad$ 1s $\quad$ 2s $\qquad\quad$ 2p

Now we're faced with a new quandry. How can we form four equivalent bonds when the four half-filled orbitals are not equivalent? We've assumed that carbon will use the same atomic orbitals that it has in the isolated atom. These may not be the best orbitals to use in bonding.

There is a mathematical way to combine the 2s orbital and the three 2p orbitals on carbon to produce four equivalent orbitals for carbon to use in bonding. When we construct these "hybrid" orbitals, we find that they have 109.5° angles between them! We call these hybrids sp^3 orbitals (s-p-three; don't confuse this superscript with the superscripts in electron configurations). Now we have:

C $\quad$ $\underline{\uparrow\downarrow}$ $\qquad$ $\underline{\uparrow\downarrow}$ $\underline{\uparrow\downarrow}$ $\underline{\uparrow\downarrow}$ $\underline{\uparrow\downarrow}$
$\qquad$ 1s $\qquad\qquad\quad$ sp^3

with the dashed arrows ($\downarrow$) representing the electrons from the four hydrogen atoms. This is to remind us that we only find carbon atoms with sp^3 hybrid orbitals when the carbon has made four bonds.

We could use sp³ hybrids to explain the bonding between the oxygen atom and the hydrogen atoms in water. These hybrids would give us a bond angle of 109.5°.

What hybrids does boron use in BF_3?

Boron's ground state electron configuration is:

B: $\underset{1s}{\underline{\uparrow\downarrow}}$ $\underset{2s}{\underline{\uparrow\downarrow}}$ $\underset{2p}{\underline{\uparrow}\ \underline{}\ \underline{}}$

We need three half-filled atomic orbitals to form three equivalent bonds. We can excite one electron from the 2s orbital to an empty 2p orbital:

B: $\underset{1s}{\underline{\uparrow\downarrow}}$ $\underset{2s}{\underline{\uparrow}}$ $\underset{2p}{\underline{\uparrow}\ \underline{\uparrow}\ \underline{}}$

Again, the 2s orbital and two 2p orbitals can be mathematically combined to give three equivalent sp² hybrid orbitals, in a plane with angles of 120° between them. Each sp² orbital can overlap a half-filled orbital on fluorine to form a σ bond.

Problem 9-2:

Describe the bonding in formaldehyde:

$$\begin{array}{c} H \\ \diagdown \\ \diagup \\ H \end{array} C = \ddot{O}$$

There are two approaches to this problem. Each begins with the Lewis structure.

Solution I:

From the Lewis structure, we predict bond angles of 120° around the carbon atom. Therefore, the carbon atom is using sp² hybrid orbitals to form σ bonds, overlapping half-filled orbitals on the hydrogen atoms and the oxygen atom. The remaining half-filled p orbital on the carbon atom is used to form a π bond with a half-filled p orbital on the oxygen atom.

Solution II:

From the Lewis structure we see that carbon makes three σ bonds and one π bond. One half-filled p orbital on the carbon atom is needed to form the π bond with oxygen. Then three sp² hybrids are formed from the remaining half-filled 2s orbital and the remaining two half-filled 2p orbitals on the carbon atom.

Problem 9-3:

Describe the bonding in i) ethylene:

$$H\diagdown \atop H \diagup C = C \diagup H \atop \diagdown H$$

and ii) acetylene:

$$H - C \equiv C - H$$

Solution:

i) From the Lewis structure of ethylene each carbon uses one half-filled p orbital to form the π bond, and sp^2 hybrid orbitals to form the σ bonds. The p orbital is perpendicular to plane containing the three sp^2 hybrid orbitals. The p orbitals on the two carbon atoms must be parallel to overlap. Therefore ethylene is a planar molecule. There is no rotation about a double bond: any rotation would destroy the side-to-side overlap of the p orbitals in the double bond.

ii) From the Lewis structure of acetylene, each carbon uses two 2p orbitals to form the two π bonds. The remaining 2s orbital and 2p orbital are combined to form two sp hybrid orbitals on each carbon. Each carbon uses one sp hybrid orbital to form a σ bond with a hydrogen atom and the other to form a σ bond with the other carbon atom.

Expanded Octets

To describe the bonding in molecules with expanded octets, we need a hybrid orbital on the central atom for each pair of electrons around the central atom in the Lewis structure. We can use a d orbital, an s orbital, and three p orbitals to form five dsp^3 hybrid orbitals. We can use two d orbitals, an s orbital, and three p orbitals to form six d^2sp^3 orbitals.

Summary of the Localized Electron Model

In the Localized Electron Model we make mathematical combinations of atomic orbitals on the same atom to form hybrid orbitals. An orbital on one atom overlaps an orbital on another atom to form a bond.

The Molecular Orbital Model

The molecular orbital approach to bonding constructs orbitals which are spread out over the entire molecule. Each molecular orbital has a particular energy. Each molecular orbital can be occupied by two electrons. Then, we can write an electron configuration for the molecule, analogous to the electron configurations we wrote for atoms.

Molecular orbitals are wave functions, solutions to the quantum mechanical equations for the molecule. These solutions are always combinations of the atomic orbitals on the separated atoms.

Some combinations of atomic orbitals are lower in energy than the separated atomic orbitals. These combinations are "bonding molecular orbitals."

Some combinations of atomic orbitals are higher in energy than the separate atomic orbitals. These molecular orbitals are "anti-bonding."

Electrons in bonding molecular orbitals, have a high probability of being found between the nuclei. Therefore, electrons in these orbitals favor bonding.

Electrons in anti-bonding molecular orbitals, on the other hand, have a high probability of being found beyond the nuclei. Electrons in anti-bonding orbitals do not reduce the repulsion between the nuclei.

Every bonding molecular orbital has a corresponding anti-bonding orbital.

Labels on molecular orbitals indicate their symmetry (σ or π). An asterisk (*) indicates that an orbital is anti-bonding. In homonuclear diatomic molecules (H_2, O_2, F_2, etc.), subscripts indicate the atomic orbitals used to form the molecular orbitals.

Rules for Forming Molecular Orbitals

1. Atomic orbitals must be of similar energies to interact to form molecular orbitals.

2. Atomic orbitals must overlap to form molecular orbitals. Inner atomic orbitals do not overlap sufficiently. Valence atomic orbitals are used to form molecular orbitals.

3. Atomic orbitals must be of the right symmetry to form molecular orbitals:

   a. An s orbital on one atom can overlap an s orbital on another atom to form a σ bond.

   b. A p orbital on one atom can overlap a p orbital on another atom end-to-end to form a σ bond.

   c. An s orbital on one atom can overlap the end of a p orbital on another atom to form a σ bond.

   d. A p orbital on one atom can overlap a p orbital on another atom side-to-side to form a π bond.

   e. An s orbital cannot interact with a p orbital "from the side": There is no net overlap.

4. Orbitals are conserved. The number of molecular orbitals formed always equals the number of atomic orbitals combined to form them.

5. For every bonding combination of atomic orbitals (bonding molecular orbitals) there is a corresponding anti-bonding combination (anti-bonding molecular orbital).

Orbital-Filling Diagrams

Orbital filling diagrams such as those in Figure 9.40 indicate the relative energies of the molecular orbitals in a molecule.

We used atomic orbitals filling diagrams to write electron configurations for atoms. Now we will use these molecular orbital filling diagrams to write electron configurations for molecules. We will use the same "filling rules" that we used for electron configurations:

a. Each molecular orbital can be occupied by two electrons.

b. Fill the lowest energy levels first.

c. When there are degenerate orbitals (orbitals of equal energy), electrons will not be paired up in any of these orbitals until each orbital is "half-filled." The electrons in these "half-filled" orbitals will have parallel spins.

Problem 9-4:

What is the electron configuration for N_2?

Solution:

a. Draw the orbital filling diagram: Include the molecular orbitals formed from the valence shell atomic orbitals in nitrogen:

___ σ_{2p}*

___ ___ π_{2p}*

___ σ_{2p}

___ ___ π_{2p}

___ σ_{2s}*

___ σ_{2s}

140

b. Place the valence electrons from the nitrogen atoms into the molecular orbitals using the filling rules. There are ten valence electrons, five from each nitrogen atom [(He)$2s^2 2p^3$].

___ $\sigma_{2p}{}^*$

___ ___ $\pi_{2p}{}^*$

↑↓ σ_{2p}

↑↓ ↑↓ π_{2p}

↑↓ $\sigma_{2s}{}^*$

↑↓ σ_{2s}

or $(\sigma_{2s})^2 (\sigma_{2s}{}^*)^2 (\pi_{2p})^4 (\sigma_{2p})^2$

N_2 has 8 bonding electrons and 2 anti-bonding electrons: 6 "net bonding electrons": a triple bond

Problem 9-5:

Does O_2 have any unpaired electrons?

Solution:

a. Write the filling diagram: Use the same diagram as that for N_2. Each oxygen atom has six valence electrons. [O: (He)$2s^2 2p^4$].

___ $\sigma_{2p}{}^*$

↑ ↑ $\pi_{2p}{}^*$

↑↓ σ_{2p}

↑↓ ↑↓ π_{2p}

↑↓ $\sigma_{2s}{}^*$

↑↓ σ_{2s}

or $(\sigma_{2s})^2 (\sigma_{2s}{}^*)^2 (\pi_{2p})^4 (\sigma_{2p})^2 (\pi_{2p}{}^*)^2$

b. O_2 has 8 bonding electrons and 4 anti-bonding electrons: 4 "net bonding electrons": a double bond

c. O_2 has two unpaired electrons. O_2 is paramagnetic.

Bond Order

The Lewis structure for nitrogen gas, $:N \equiv N:$, predicts a "triple bond" in N_2; the Lewis structure for oxygen gas, $:O = O:$, predicts a "double bond" for O_2. How do we interpret the molecular orbital electron configurations for these molecules?

Remember, electrons in bonding molecular orbitals have a high probability of being found between the nuclei, drawing the nuclei together. Electrons in anti-bonding molecular orbitals have a high probability of being found beyond the nuclei, drawing the nuclei apart.

A pair of electrons in a bonding molecular orbital is equivalent to a bonding pair of electrons in the Lewis model. A pair of electrons in an anti-bonding molecular orbital "cancels" the bonding effect of a pair of electrons in a bonding molecular orbital.

"Bond order" is the net number of pairs of bonding electrons. Bond order = (bonding electrons − non-bonding electrons)/2. Therefore, the bond order in N_2 is:

$$\frac{8 - 2}{2} = 3;$$

and the bond order in O_2 is:

$$\frac{8 - 4}{2} = 2.$$

These bond orders match the bonds predicted by the Lewis structures.

Problem 9-6:

$O_2{}^+$ has a stronger, shorter bond than O_2.

Does $N_2{}^+$ have a stronger, shorter bond than N_2?

Solution:

a. Write the orbital filling diagrams for $N_2{}^+$ and N_2:

N$_2$ has ten valence electrons: each N is $(He)2s^2 2p^3$;

$N_2{}^+$ has nine valence electrons, one less than N_2:

N_2:
          ___ σ_{2p}*

     ___ ___ π_{2p}*

     ↑↓ σ_{2p}

  ↑↓ ↑↓ π_{2p}

    ↑↓ σ_{2s}*

    ↑↓ σ_{2s}

$N_2{}^+$:
          ___ σ_{2p}*

     ___ ___ π_{2p}*

     ↑ σ_{2p}

  ↑↓ ↑↓ π_{2p}

    ↑↓ σ_{2s}*

    ↑↓ σ_{2s}

b. Determine the bond orders:

N_2: $\dfrac{8-2}{2} = 3$

$N_2{}^+$: $\dfrac{7-2}{2} = 2.5$

c. N_2 has a <u>triple</u> bond; $N_2{}^+$ has a "two-and-a-half" bond. The triple bond in N_2 is shorter and stronger than the bond in $N_2{}^+$.

Let's compare the Localized Electron Model of bonding with the Molecular Orbital Model of bonding. The Localized Electron Model is easy to use. We can predict relative bond lengths and bond strengths from Lewis structures. We can predict a molecule's geometry and polarity using VSEPR. However, this model can't describe molecules like SO_2 or SO_3 without using resonance structures. The Localized Electron Model is limited to describing molecules and ions with pairs of electrons, not species with an odd number of electrons, like NO.

The Molecular Orbital Model of bonding can predict relative bond lengths and strengths and magnetic properties for diatomic molecules easily, including those with an odd number of electrons. However, there is no easy way to apply the Molecular Orbital Model to complex molecules.

We can combine these models, using the Localized Electron Model to describe the σ bonding system in a molecule, and the Molecular Orbital Model to describe the π bonding system in a molecule. Then we can describe the bonding in SO_2 with localized σ bonds between sulfur and each of the oxygen atoms, using sp^2 hybrids on the sulfur atom. The π bonding system is described by a bonding molecular orbital which combines a p orbital on each oxygen atom with the p orbital on the sulfur atom. These p orbitals overlap side-to-side to give a π bonding molecular orbital which extends over the whole molecule.

Each model has its own advantages.

9-1.   Describe the bond angles in the following molecules:

a. CH_2Cl_2

b. 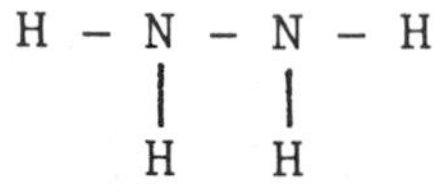   H – N – N – H
             | |
             H   H

c. 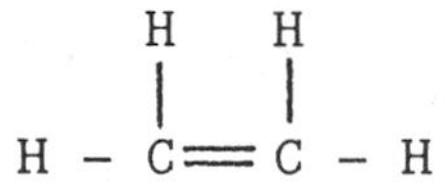   H   H
             | |
       H – C = C – H

d.   H – Be – H

9-2.   Describe the atomic orbitals or hybrid orbitals used by each atom in the molecules in 9-1.

9-3.   What hybrid orbitals does the central atom use in bonding in the following molecules and ions?

a. BF_3 e. SF_6

b. NH_3 f. PCl_5

c. $CHCl_3$ g. CO_2

d. NO_3^-

9-4.   What hybrid orbitals does the indicated atom use in the following molecules?

a.   oxygen in H_2O_2 [H – Ö – Ö – H]

b.   carbon in   H – C ≡ N:

                                   H
                                   |
c.   carbon in   H – C = Ö

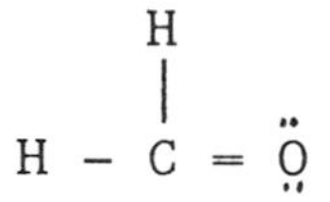

d.   sulfur in   :Ö – S = Ö

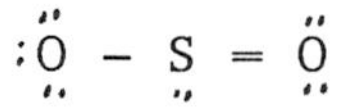

9-5.   Use the orbital filling diagram in Figure 9.40 to write electron configurations for the following diatomic molecules and ions:

a. O_2 d. O_2^+

b. N_2 e. O_2^-

c. F_2 f. Be_2

9-6.   For each of the species in 9-6, determine

   i)   The bond order

   ii)   the number of unpaired electrons

9-7.   We can use the molecular orbital filling diagram for Row 2 homonuclear diatomic molecules for Row 2 heteronuclear diatomic molecules.

   Write the electron configuration for the following diatomic molecules and ions:

   a.   NO

   b. NO^+

   c.   CN

   d.   OF

9-8.   For each of the species in 9-7, determine

   i)   the bond order

   ii)   the number of unpaired electrons

CHAPTER NINE: SELF-TEST ANSWERS

9-1. a. All angles are 109.5°

b. The H – N – H bond angles and the H – N – N bond angles are approximately 107°.

c. The H – C – H bond angles and the H – C – C bond angles are 120°.

d. The H – Be – H bond angle is 180°.

9-2. a. The C uses sp³ hybrids;

Each H uses 1s orbitals;

It's hard to know what hybrids Cl uses, as we have no bond angles to look at; we might propose sp³ hybrids.

b. Each N uses sp³ hybrids

Each H uses a 1s orbital

c. Each C uses sp² hybrids for the σ – bonds and a p orbital for the π bond.

Each H uses a 1s orbital

d. Be uses sp hybrids

Each H uses a 1s orbital.

9-3. a. sp² hybrids

b. sp³ hybrids

c. sp³ hybrids

d. sp² hybrids to form σ bonds
a p orbital to form π bonds

e. d²sp³ hybrids

f. dsp³ hybrids

g. sp hybrids to form σ bonds
p orbitals to form π bonds

9-4. a. sp³

b. sp

c. sp²

d. sp²

9-5. 9-6.			σ_{2s}	$\sigma_{2s}*$	π_{2p}	σ_{2p}	$\pi_{2p}*$	$\sigma_{2p}*$	Bond order	Unpaired e^-'s
	a.	O_2	2	2	4	2	2		2	2
	b.	N_2	2	2	4	2			3	0
	c.	F_2	2	2	4	2	4		1	0
	d.	O_2^+	2	2	4	2	1		2 1/2	1
	e.	O_2^-	2	2	4	2	3		1 1/2	1
	f.	Be_2	2	2					0	0

9-7. 9-8.			σ_{2s}	$\sigma_{2s}*$	π_{2p}	σ_{2p}	$\pi_{2p}*$	$\sigma_{2p}*$	Bond order	Unpaired e^-'s
	a.	NO	2	2	4	2	1		2 1/2	1
	b.	NO^+	2	2	4	2			3	0
	c.	CN^-	2	2	4	2			3	0
	d.	OF	2	2	4	2	3		1 1/2	1

CHAPTER TEN: LIQUIDS AND SOLIDS

Solids have a fixed volume and shape. Liquids have a fixed volume, but no fixed shape. Solids melt to form liquids. Liquids vaporize to form gases. Gases are very compressible, while liquids are nearly incompressible. What forces hold molecules together in a liquid or solid? What forces keep molecules apart in a liquid or solid when we try to compress them? What forces are there between particles (molecules, atoms, or ions) in solids? In liquids? Why is methane (CH_4) a gas while water (H_2O) is a liquid?

We used a microscopic model for atoms and molecules in a gas to explain and predict "macroscopic behavior" of gases. We didn't need to give much attention to the forces between molecules in a gas. Under "ordinary" conditions, gas molecules are far enough apart from each other that the inter-molecular forces are weak enough to be ignored.

Problem 10-1:

Gaseous water at 400°C and 1 atm pressure has a density of $3.26 \times 10^{-3} g/cm^3$. Liquid water at 25°C and 1 atm pressure has a density of .997 g/cm^3.

    i)   What volume of gaseous water at 400° and 1 atm pressure contains one gram of water?

    ii)   What volume of liquid water at 25° and 1 atm pressure contains one gram of water?

Solution:

    i) 3.26×10^{-3} g of gaseous water occupy 1.00 cm^3: the volume occupied by one gram of gaseous water is:

$$\frac{1.00 \text{ cm}^3}{3.26 \times 10^{-3} \text{ g}} = \frac{307 \text{ cm}^3}{\text{g}}$$

    ii)   .997 g of liquid water occupy 1.00 cm^3:

$$\frac{1.00 \text{ cm}^3}{.997 \text{ g}} = \frac{1.003 \text{ cm}^3}{\text{g}}$$

From this calculation we can see that the molecules of water must be very close together in the liquid.

Intermolecular Forces

We've seen the affect of ionic size and charge on the lattice energy of an ionic solid. The electrostatic attractions between ions of opposite charge in these solids are responsible for the hard brittle nature of ionic solids, and their high melting points. What forces keep molecules together in molec-ular solids and liquids?

Dipole-Dipole Forces

A polar molecule acts as a dipole, with a negative end and a positive end. The positive end of one dipole is attracted to the negative end of another. The positive ends of these dipoles repel each other. In a collection of polar molecules, the dipoles will arrange themselves to maximize the attractions between the molecules.

In Chapter 8 we learned to predict the polarity of molecules. First we used the Lewis structure for the molecule to predict the molecule's geometry. Then we used the difference in electronegativities of atoms in the bonds to decide whether the bonds are polar. Finally, we combined the bond polarity with the molecular geometry to predict whether the molecule is polar.

Hydrogen Bonds

Ammonia (NH_3), water (H_2O), and hydrogen fluoride (HF) each have higher boiling points than expected from trends for hydrides. (See Figure 10.4 in the text.) The greater the intermolecular forces between molecules in a liquid, the higher the boiling of the liquid. There must be a stronger interaction between these molecules than an ordinary dipole-dipole interaction. We call this unique interaction a hydrogen bond.

In order for a hydrogen bond to form between two atoms the hydrogen atom must be covalently bound to a small electronegative atom (N, O, or F) and there must be another small electronegative atom with a "lone pair" of electrons to form the bond. The strongest hydrogen bonds form when the three atoms lie in a straight line. The hydrogen bonds in water can be represented as: —O-H---:O-, where the solid line (-) represents the covalent bond, and the dashed line (---) represents the hydrogen bond.

In ice, each water molecule makes 4 hydrogen bonds with other water molecules.

Hydrogen bonds are stronger than dipole-dipole forces.

London Dispersion Forces

Oxygen, O_2, and nitrogen, N_2, can be liquified under pressure. What attractive forces are there between these non-polar molecules? Since these gases have very low boiling points, the attractive forces between these molecules must be weaker than dipole-dipole forces.

Fritz London proposed that these forces arise from a momentary imbalance in the symmetric electron distribution of the non-polar molecule. This momentary dipole (transient dipole) can "induce" a transient dipole in a neighboring molecule. These weak transient dipoles attract each other. London dispersion forces exist between all molecules, but are overpowered by stronger dipole-dipole interactions or hydrogen bonds. London dispersion forces depend on how easily the electron distribution of a molecule can be distorted. This distortion is called "polarization." In general, the farther the outer electrons are from the nucleus, the easier it is to distort the electron distribution, the greater the polarizability, and the greater the London forces between the molecules.

Properties of Liquids

The boiling point of a liquid and its surface tension depend on the strength of the intermolecular forces - the stronger the forces, the higher the boiling point and the higher the surface tension.

The viscosity of a liquid depends on both the strength of the intermolecular forces, and the shape of the molecule. Liquids with large complicated molecules are more viscous because their molecules can become entangled.

Problem 10-2:

Arrange the following in order of increasing boiling point:

$$CF_4, \quad CH_2F_2, \quad H_2O, \quad He, \quad Ar, \quad CH_3OH \text{ (methanol)}$$

Explain your answer.

The stronger the intermolecular forces, the higher the boiling point.

Solution:

First, separate these liquids by the kinds of forces you expect to find between the molecules.

Water and methanol can each form hydrogen bonds. CF_4 and CH_2F_2 each have polar bonds, but CF_4 is non-polar. CH_2F_2 is a polar molecule, with dipole-dipole forces between its molecules in the liquid. CF_4, He, and Ar are non-polar molecules.

Tentatively, we could group these molecules:

$$CF_4, \text{ He, Ar} < CH_2F_2 < H_2O, \text{ } CH_3OH$$

In CF_4, He and Ar there are London dispersion forces between the molecules. London dispersion forces tend to increase with increasing molecular weight. The boiling points for the non-polar molecules should be in the order:

$$\text{He} < \text{Ar} < CF_4$$

Because CH_2F_2 is polar, its boiling point should be higher than that of CF_4. Although methanol and water each form hydrogen bonds, each water molecule can form <u>four</u> hydrogen bonds. Each methanol molecule forms only <u>two</u> hydrogen bonds on the average.

The final order of boiling points we'd predict is:

$$\text{He} < \text{Ar} < CF_4 < CH_2F_2 < CH_3OH < H_2O$$

150

Solids

Crystalline solids include ionic solids, molecular solids, network solids and metals. In each solid there is a regular arrangement of ions, atoms or molecules. These crystal structures are studied by x-ray analysis. X-ray analysis then yields the smallest repeating unit in the lattice (the unit cell) and the arrangement of the atoms within the unit cell. Figures 10.17 through 10.20 show some simple unit cells.

We can use the x-ray data to calculate the size of atoms.

For example:

Problem 10-3:

Metallic gold crystallizes in a face-centered cubic lattice, with a unit cell 4.07×10^{-10} m on a side. If we think of gold atoms as hard spheres touching along the diagonal of the face of the unit cell, what is the radius of the gold atom? (See Figure 10.20c in the text.)

Solution:

Use the Pythagorean theorem to calculate the diagonal of the face of the unit cell from the length of the edge of the unit cell:

$$(\text{diagonal})^2 = (\text{edge})^2 + (\text{edge})^2$$

$$d^2 = 2(4.07 \times 10^{-10} \text{ m})^2$$

$$d = \text{diagonal} = 5.75 \times 10^{-10} \text{ m}$$

The diagonal of the face equals the radius of a gold atom at one corner (r), plus the diameter of the atom in the center of the face (2r), plus the radius of the atom at the opposite corner (r).

$$d = r + 2r + r$$

$$r = \frac{d}{4} = \frac{5.75 \times 10^{-10} \text{ m}}{4} = 1.44 \times 10^{-10} \text{ m}$$

Network (Atomic) Solids

These solids have networks of covalent bonds. Solids with a three-dimensional network, like diamond and SiO_2, are hard, brittle and high melting. Graphite has planes of covalently bonded carbon atoms. These planes have weak interactions with each other. These weak interactions allow the layers to slide over each other.

Metallic Sounds

There are no directional bonds in metallic solids. Therefore, metals can be hammered or rolled into sheets, or pulled into wires, without breaking bonds. One model for the bonding in metals uses a molecular orbital approach, combining the valence s and p orbitals on each metal atom to form many closely spaced molecular orbitals which are delocalized over the entire crystal. The valence electrons occupy these molecular orbitals. Because the molecular orbitals are very closely spaced and are only partially occupied, electrons in filled MO's can easily be excited into empty MO's to carry current through the metal.

Ionic Solids

There are no directional bonds in ionic solids either. But one layer of ions can't slide easily over another layer of ions because of the strong attraction between cations and anions in the planes, and the repulsions between ions of like charge.

Vapor Pressures

In the kinetic molecular theory of gases (in Chapter 5) we saw that kinetic energy is distributed among the molecules (see Figures 5.14 and 5.15) in a gas. This same distribution of kinetic energies among molecules occurs in liquids and in solids.

Why do liquids in open containers evaporate? Some of the molecules in the liquid have enough kinetic energy to overcome the attractive forces between the molecules holding them in the liquid. When these molecules with higher kinetic energy escape, the average kinetic energy of the remaining molecules is lower. Thus, the temperature of the remaining liquid drops as liquid evaporates.

Vaporization is favored by high temperature (so that more molecules have sufficient kinetic energy to escape the liquid) and small attractive forces between the molecules.

What happens when we place a liquid in a <u>closed</u> container? At first, there is no vapor above the liquid. Some of the liquid molecules have enough kinetic energy to escape into the gas phase. The molecules in the gas phase have a distribution of kinetic energies; some of these have a low enough kinetic energy to be trapped when they collide with the surface of the liquid. Soon there is an equilibrium, a balance between molecules escaping from the liquid and molecules returning to the liquid. The equilibrium pressure of the gas (vapor) above the liquid is the liquid's <u>vapor pressure and temperature</u>.

The vapor pressure of a liquid increases with increasing temperature. The temperature dependence follows the equation:

$$P_{vapor} = \text{constant} \cdot e^{\dfrac{-\Delta H_{vaporization}}{RT}}$$

or:

$$\ln p_{vapor} = -\frac{\Delta H_{vapor'n}}{RT} + C$$

$$\text{or} \quad \ln\left(\frac{P_{2vapor}}{P_{1vapor}}\right) = -\frac{\Delta H_{vapor'n}}{R}\left(\frac{1}{T_2} - \frac{1}{T_1}\right)$$

where $\Delta H_{vapor'n}$ is the enthalpy of vaporization. The enthalpy of vaporization is always positive. Thus, the vapor pressure always increases with increasing temperature.

(Optional)
Problem 10-4:

The vapor pressure of water is 187.5 torr at 65°C and 433.6 torr at 85°C. What is the enthalpy of vaporization of water?

(ln indicates the natural logarithm, or logarithm to the base e. Find the ln key on your calculator.)

$$\ln\left(\frac{433.6 \text{ torr}}{187.5 \text{ torr}}\right) = -\frac{\Delta H}{R}\left(\frac{1}{(273 + 85)K} - \frac{1}{(273 + 65)K}\right)$$

We need to use R in units of $\frac{J}{K \text{ mol}}$: $R = \frac{8.31441 \text{ J}}{K \text{ mol}}$

$$\ln 2.312 = \frac{-\Delta H}{\frac{8.31441 \text{ J}}{K \text{ mol}}}\left(\frac{1}{358} - \frac{1}{338}\right)$$

$$.8381 = \frac{-\Delta H}{\frac{8.31441 \text{ J}}{K \text{ mol}}}\left(\frac{.00279}{K} - \frac{.00296}{K}\right)$$

Solving for ΔH: $\quad \Delta H = .8381\left(\frac{8.3144 \text{ J}}{K \text{ mol}}\right)\left(\frac{1}{.00017} K\right) = \frac{41 \text{ kJ}}{\text{mol}}$

This is the heat required to vaporize one mole of liquid water.

Make sure you have a qualitative understanding of how vapor pressure varies with temperature:

i)   The vapor pressure of every liquid increases with temperature.

ii) Liquids with stronger intermolecular forces tend to have a lower vapor pressure at a given temperature.

iii) Liquids with stronger intermolecular forces have larger enthalpies of vaporization.

153

Phase Changes

Look at the heating curve of water (Figure 10.46) in the text. As energy is added to solid ice (from A to B on the curve) the average kinetic energy of the molecules vibrating in the solid increases. From point B to point C the temperature remains constant as the ice melts to form liquid water. The energy added at this point must be increasing the potential energy of the molecules as some of the hydrogen bonds are broken. The average kinetic energy of the molecules in the liquid water is the same as that of the molecules in the ice at this temperature.

From point C to point D, the average kinetic energy of the molecules in the liquid increases. The molecules in the liquid vibrate and tumble.

From point D to point E the temperature remains constant as the liquid vaporizes. Again, during this phase change, the average kinetic energy of the molecules remains the same. The energy added must overcome nearly all of the remaining attractive forces between the water molecules.

If we assume that the heat to melt the ice (the heat of fusion) and the heat to vaporize the liquid water (the heat of vaporization) depend on the number of hydrogen bonds broken, we can estimate that only a small fraction of the hydrogen bonds is broken when ice melts. The remaining hydrogen bonds are broken when the liquid water vaporizes.

CHAPTER TEN: SELF-TEST

10-1. Describe the forces between particles in each of the following.

a. $NaCl(s)$

b. $N_2(\ell)$

c. $Ag(s)$

d. $HF(\ell)$

e. $NH_3(\ell)$

f. $I_2(s)$

g. $CCl_4(\ell)$

h. $HCl(\ell)$

i. C (diamond)

10-2. In each pair, choose the substance you expect to have the higher boiling point.

a. Ne or Ar

b. $CH_3\text{-}O\text{-}CH_3$ or CH_3CH_2OH

c. $CH_3CH_2CH_2CH_3$ or $CH_3CH_2CH_2CH_2CH_3$

d. HCl or HF

e. CF_4 or CH_2Cl_2

10-3. In each pair, choose the liquid you expect to have the higher vapor pressure.

a. $CH_3\text{-}O\text{-}CH_3$ or CH_3CH_2OH

b. $CH_3CH_2CH_2CH_3$ or $CH_3CH_2CH_2CH_2CH_3$

c. CH_3OH or H_2O

10-4. Uranium crystallizes in a body centered cubic lattice, with one atom at each corner of the cube, and one atom in the center.

a. How much of the atom at the corner belongs to the unit cell?

b. How many atoms are in the unit cell?

c. How many neighbors does each atom "touch"?

d. If the edge of the unit cell is 3.45×10^{-8} cm, what is the volume of the unit cell?

e. What volume would one mole of uranium atoms occupy?

f. What is the density of uranium? (Hint: Use the result from d.)

10-5. At sea level, liquid A boils at 80°C. Liquid B boils at 95°C. At 95°C the vapor pressure of liquid C is 700 torr.

a. Which liquid has the strongest intermolecular forces?

b. Which liquid has the weakest intermolecular forces?

10-6. Look at the heating curve of water (Figure 10.46 in the text).

a. Do the water molecules at Point C have a higher or lower average kinetic energy than the molecules at point B?

b. Do the water molecules at Point C have a higher or lower average potential energy than the molecules at Point B?

c. Do the water molecules at point C have a higher or lower average kinetic energy than the molecules at Point D?

10-7. (Optional)

The heat of vaporization of ethanol is 39.3 kJ/mole. Ethanol boils at 78.5°C (352.7 K); that is, at 78.5°C, the vapor pressure of ethanol is 760 mm Hg. At what temperature is the vapor pressure of ethanol 625 mm Hg?

CHAPTER TEN: SELF-TEST ANSWERS

10-1.   a. Attractions between cations and anions; repulsions between ions of like charge.

   b. London dispersion forces

   c. There are no specific forces between silver atoms; the metal is held together by a "sea" of loosely held electrons in closely spaced bonding molecular orbitals.

   d. hydrogen bonds

   e. hydrogen bonds

   f. London dispersion forces

   g. London dispersion forces (CCl_4 is non-polar.)

   h. dipole-dipole forces

   i. covalent bonds

10-2.   a. Ar; heavier atoms have stronger London dispersion forces.

   b. CH_3CH_2OH; ethanol forms hydrogen bonds.

   c. $CH_3CH_2CH_2CH_2CH_3$; heavier molecules have stronger London dispersion forces.

   d. CH_2Cl_2 is polar; CF_4 is non-polar; dipole-dipole forces are stronger than London dispersion forces.

10-3.   The stronger the intermolecular forces, the <u>lower</u> the vapor pressure.

   a. CH_3-O-CH_3; ethanol forms hydrogen bonds

   b. $CH_3CH_2CH_2CH_3$

   c. CH_3OH

10-4.   a. 1/8 th

   b. 8 atoms at corners × 1/8 + 1 atom in the center = 2 atoms/unit cell

   c. Each atom touches <u>8</u> others.

   d. 4.11×10^{-23} cm^3 is the volume of the unit cell.

   e. $\dfrac{4.10 \times 10^{-23} \text{ cm}^3}{2 \text{ atoms}} \times \dfrac{(6.022 \times 10^{23} \text{ atoms})}{\text{mole}} = \dfrac{12.4 \text{ cm}^3}{\text{mole}}$

   f. $\dfrac{238 \text{ g/mole}}{12.4 \text{ cm}^3/\text{mole}} = \dfrac{19.2 \text{ g}}{\text{cm}^3}$

157

10-5. At any temperature, the vapor pressure of A is the largest, and the vapor pressure of C is the smallest;

    a. A has the weakest intermolecular forces.

    b. C has the strongest intermolecular forces.

10-6. a. Molecules at C and B have the _same_ average kinetic energy.

    b. Molecules at C have a _higher_ average potential energy than those at B.

    c. Molecules at C have a _lower_ average kinetic energy than those at D.

10-7. $\ln \dfrac{P_2}{P_1} = \dfrac{-\Delta H}{R} \left(\dfrac{1}{T_2} - \dfrac{1}{T_1} \right)$

$$\ln \left(\frac{760}{625} \right) = \frac{-39.3 \times 10^3 \text{ J/mole}}{8.3144 \text{ J/K mole}} \left(\frac{1}{352.7} - \frac{1}{T_1} \right)$$

$0.1956 = -4727 \, (0.002835 - 1/T_1)$

$-4.138 \times 10^{-5} = +2.835 \times 10^{-3} - 1/T_1$

$1/T_1 = 2.876 \times 10^{-3} \text{ K}^{-1}$

$T_1 = 347.7 \text{ K} = 74.5°C$

CHAPTER ELEVEN: SOLUTIONS

Solution Composition

Solutions are homogeneous mixtures, that is, their composition is uniform throughout. We can describe the composition of a solution by the amount of solute dissolved in a given amount of solvent or in a given amount of solution. The amounts of solute, solvent and solution can each be given in moles, mass, or volume. The units for these concentrations each depend on how the amounts of solute and solvent or solution are measured.

Concentration	Amount of Solute	Amount of Solvent	Amount of Solution	Units
Mass %	Grams		100 Grams	% has no units
molarity	moles	--	liters	moles/L = M
molality	moles	kg	--	moles/kg = m
mole fraction	moles	--	moles solute + moles solvent	X (X has no units)
normality	equivalents*	--	liters	equiv/L = N

*The definition of an equivalent depends on the reaction. For acid-base reactions, one equivalent will produce or consume one mole of hydrogen ions. One mole of sulfuric acid, H_2SO_4, is two equivalents if it reacts completely with a strong base; one mole of barium hydroxide, $Ba(OH)_2$, is two equivalents.

For redox reactions, one equivalent will produce or consume one mole of electrons. One mole of $KMnO_4$ is five equivalents when the reduction half reaction is:

$$MnO_4^- + 8 \ H^+ + 5 \ e^- \rightarrow Mn^{2+} + 4 \ H_2O$$

One mole of $KMnO_4$ is three equivalents when the reduction half reaction is:

$$MnO_4^- + 4 \ H^+ + 3 \ e^- \rightarrow MnO_2(s) + 2 \ H_2O$$

Problem 11-1:

15.0 g of toluene (C_7H_8) are dissolved in 180.0 g of benzene (C_6H_6) to form 221.0 mL of solution.

Calculate the concentration of toluene in this solution in:

i) mass percent

ii) molarity

iii) molality

iv) mole fraction

Solution:

i) mass percent $= \dfrac{\text{grams toluene}}{\text{grams solution}} \times 100$

$= \dfrac{15.0 \text{ g}}{15.0 \text{ g} + 180.0 \text{ g}} \times 100 = 7.69\%$

ii) molarity $= \dfrac{\text{moles toluene}}{\text{liters solution}}$

The molecular weight of C_7H_8 is: 92.14 g/mole

moles toluene $= \dfrac{15.0 \text{ g}}{92.14 \text{ g/mole}} = 0.163 \text{ moles}$

$221.0 \text{ mL} \times \dfrac{1 \text{ L}}{1000 \text{ mL}} = 0.2210 \text{ L}$

$\text{molarity}_{toluene} = \dfrac{0.163 \text{ moles}}{0.2210 \text{ L}} = 0.738 \text{ M}$

iii) molality $= \dfrac{\text{moles toluene}}{\text{kg solvent}}$

moles toluene = 0.163 moles (from ii)

$180.0 \text{ g benzene} \times \dfrac{1 \text{ kg}}{1000 \text{ g}} = 0.1800 \text{ kg benzene}$

$\text{molality}_{toluene} = \dfrac{0.163 \text{ moles toluene}}{0.1800 \text{ kg benzene}} = 0.906 \text{ m}$

iv) mole fraction$_{toluene} = \dfrac{\text{moles toluene}}{\text{moles toluene} + \text{moles benzene}}$

moles toluene = 0.163 moles (from ii)

The molecular weight of benzene (C_6H_6) is: 78.11 g/mole

moles benzene $= \dfrac{180.0 \text{ g}}{78.11 \text{ g/mole}} = 2.304 \text{ moles}$

mole fraction$_{toluene} = \dfrac{0.163}{2.304 + 0.163} = 0.0661$

Notes:

Molarity is commonly used in chemistry laboratories because we can then easily obtain the desired amount of the solute by measuring a volume of solution. However, because the volume of a solution changes with temperature, the molarity of a solution depends on its temperature.

The mass percent, molality and mole fraction of a solute do not change with temperature. We can easily obtain the desired amount of the solute by using a measured weight of solution, and the mass percent. Solutions of a desired molality or mole fraction are easy to make. However, it is difficult to use either of these concentrations to obtain a desired amount of solute. Molality and mole fraction are often the concentration units used when we study properties of the solution itself.

The sum of all the mole fractions of the components in a solution must equal 1.00. In this solution there are only two components. As $X_{C_7H_6}$ = 0.0672, $X_{C_6H_6}$ = 1 − 0.0672 = 0.9328.

Solution Formation

The natural tendency for systems to become more disordered favors the formation of a solution. However, if forming the solution requires too much energy, the process won't be favored.

Because energy and enthalpy are state functions, we can break the process of dissolving a solute into individual hypothetical steps.

1. First, separate the particles in the solute.

2. Next, create "holes" in the solvent.

3. Finally, place the solute particles into the holes in the solvent and allow the solute and solvent molecules to interact.

The enthalpy of solution will be the sum of the change in enthalpy for these three steps.

$$\Delta H_{solution} = \Delta H_1 + \Delta H_2 + \Delta H_3$$

Imagine trying to dissolve a non-polar molecule like n-pentane ($CH_3CH_2-CH_2CH_2CH_3$) in water.

Step 1.   There are weak London dispersion forces to overcome to separate n-pentane molecules:

ΔH_1 is positive, but small

Step 2.   We must break hydrogen bonds to create "holes" in water. Hydrogen bonds are much stronger than London dispersion forces.

ΔH_2 is positive and large.

161

Step 3.   We place n-pentane molecules into the holes. n-pentane cannot form hydrogen bonds to water; there are only relatively weak London dispersion forces between n-pentane molecules and water molecules.

$$\Delta H_3 \text{ is negative, but small.}$$

Therefore, $\Delta H_{solution}$ is positive and large. n-pentane is not very soluble in water.

Imagine dissolving NaCl in water.

Step 1:   There are strong electrostatic forces between the ions. ΔH_1 is the lattice energy. ΔH_1 is positive, and very large.

Step 2:   We must break hydrogen bonds to create the holes in water. ΔH_2 is positive and large.

Step 3:   We place the sodium (Na^+) ions and chloride (Cl^-) ions in the holes. There is a very strong interaction between the polar water molecules and the ions. ΔH_3 is negative and large.

$$\Delta H_{solution\ NaCl} = \Delta H_1 + \Delta H_2 + \Delta H_3 = 3 \text{ kJ/mol}$$

$\Delta H_{solution}$ is small enough that the natural tendency towards "disorder" wins, and NaCl dissolves in water. Look at Table 11.3 for a generic view of the solution process.

If the enthalpy of solution is positive (the process is endothermic), the solubility of the solute increases with increasing temperature. If the enthalpy of solution is negative, the solubility of the solute decreases with increasing temperature.

Solubility Rules

Some salts dissolve readily in water. Other salts are nearly insoluble. It is easier to remember solubility rules for salts than to try to predict solubility from lattice energies and hydration energies (the enthalpy of hydration is the change in enthalpy when ions are dissolved in water; the enthalpy of hydration, $\Delta H_{hydration}$, is the sum of ΔH_2 and ΔH_3 above).

Henry's Law

The solubility of a gas in a liquid is proportional to the pressure of the gas above the solution; or the pressure of the gas above the solution is proportional to the concentration of the dissolved gas.

$$P_{gas} = k \cdot \text{concentration}_{gas}$$

where k depends on the solution, on the concentration units used for the gas, the units of pressure, and the temperature.

Problem 11-2:

The partial pressure of oxygen in air at sea level is 0.20 atm. The concentration of dissolved oxygen in water saturated with air at 20°C at sea level is 2.7×10^{-4} moles/liter. Calculate the Henry's law constant for oxygen in water.

Solution:

$$P_{O_2} = k_{O_2} \cdot M_{O_2}$$

$$0.20 \text{ atm} = k_{O_2} \left(\frac{2.70 \times 10^{-4} \text{ moles}}{\text{liter}} \right)$$

Solve for k:

$$k_{O_2} = \frac{0.20 \text{ atm}}{2.7 \times 10^{-4}/\text{liter}} = \frac{7.4 \times 10^2 \text{ L atm}}{\text{moles}}$$

Problem 11-3:

The partial pressure of nitrogen in air at sea level is 0.80 atm. The Henry's Law constant for nitrogen is 1.5×10^3 L atm/moles at 20°C. Calculate the concentration of nitrogen dissolved in water saturated with air at sea level.

Solution:

$$P_{N_2} = k_{N_2} \cdot M_{N_2}$$

$$0.80 \text{ atm} = \frac{1.5 \times 10^3 \text{ L atm}}{\text{mol}} \cdot M_{N_2}$$

Solve for M_{N_2}:

$$M_{N_2} = \frac{0.80 \text{ atm}}{\dfrac{1.5 \times 10^3 \text{ L atm}}{\text{mol}}} = 5.3 \times 10^{-4} \ M$$

The solubility of gases decreases with increasing temperature.

Vapor Pressure of Solutions: Non-volatile Solutes

In Chapter 10, we saw that the equilibrium vapor pressure of a liquid is due to a dynamic equilibrium, a balance between the molecules escaping into the gas from the liquid, and the molecules returning to the liquid from the gas. Now, let us add a non-volatile solute to the pure liquid.

The rate of solvent molecules escaping from the liquid is proportional to the fraction of solvent molecules in the solution. Adding the non-volatile solute reduces the rate that solvent molecules escape.

The rate that solvent molecules return to the solution from the vapor is proportional to the concentration of solvent molecules in the vapor.

When we add the non-volatile solute to the pure solvent we disturb the dynamic balance between solvent molecules escaping from the liquid and solvent molecules returning to the liquid. For a time, solvent molecules will return to the liquid faster than solvent molecules escape from the liquid. Slowly the concentration of molecules in the vapor will decrease, until a new equilibrium is reached. The equilibrium vapor pressure of this solution is lower than the vapor pressure of the pure solvent.

Many solutions obey Raoult's law:

$$P_{soln} = X_{solvent} \cdot P^{\circ}_{solvent}$$

where the vapor pressure of the solvent in the solution (P_{soln}) equals the vapor pressure of the pure solvent ($P^{\circ}_{solvent}$) times the mole fraction of the solvent ($X_{solvent}$) in the solution.

Problem 11-4:

The equilibrium vapor pressure of water at 100°C is 760.0 torr. What is the equilibrium vapor pressure of a solution of 150 grams of sucrose ($C_{12}H_{22}O_{11}$) in 845 grams of water at 100°C?

Solution:

$$P_{soln} = X_{H_2O} \cdot P^{\circ}_{H_2O}$$

$$X_{H_2O} = \frac{moles\ H_2O}{moles\ H_2O + moles\ sucrose}$$

$$moles\ sucrose = \frac{150\ g}{342.2\ g/mole} = .438\ moles$$

$$moles\ H_2O = \frac{845\ g}{18.015\ g/mole} = 46.9\ moles$$

$$X_{H_2O} = \frac{46.9\ moles\ H_2O}{46.9\ moles\ H_2O + .430\ moles\ sucrose} = \frac{46.9}{47.3} = .992$$

$$P_{soln} = .992\ (760\ torr)$$

$$= 754\ Torr$$

Adding sucrose to the water has lowered the vapor pressure of water, although not very much.

Vapor Pressure of Solutions: Volatile Solutes

For liquid-liquid solutions (liquid solutions with volatile solutes), the vapor above the solution contains both solvent and solute molecules. We can write Raoult's law for each component:

$$P_A = X_A \cdot P_A^\circ$$

$$P_B = X_B \cdot P_B^\circ$$

where P_A and P_B, the vapor pressures of component A and B above the solution, X_A and X_B are mole fractions of A and B in the solution, and P_A° and P_B° are the equilibrium vapor pressures of pure A and pure B.

Then the vapor pressure of the solution is:

$$P = P_A + P_B = X_A P_A^\circ + X_B P_B^\circ$$

Solutions which obey Raoult's law are "ideal solutions." Raoult's law presumes that only dilution of the solvent and solute molecules affects the vapor pressure of the solution.

Deviations from Raoult's Law

If the interactions between solvent molecules and solute molecules are much stronger in the solution than in the pure liquids, both solvent and solute molecules will escape from the solution at lower rates than predicted by their mole fractions. The vapor pressure of the solution is lower than the vapor pressure predicted by Raoult's law. The enthalpy of solution will be negative if the interactions between solvent and solute molecules in solution are stronger than those in the pure solvent and the pure solute. Then we expect to see negative deviations from Raoult's law.

Problem 11-5:

What is the equilibrium vapor pressure of a solution containing 36.2 g of toluene (C_7H_8) and 80.6 g of benzene (C_6H_6) at 79.0°C. At 79.0°C, the vapor pressure of pure toluene is 290 torr; the vapor pressure of pure benzene is 745 torr. Assume the solution obeys Raoult's law.

Solution:

Use Raoult's law:

$$P = P_{tol} + P_{benz} = X_{tol} P_{tol}^\circ + X_{benz} P_{benz}^\circ$$

$$X_{tol} = \frac{moles_{tol}}{moles_{tol} + moles_{benz}}$$

$$X_{benz} = \frac{moles_{benz}}{moles_{tol} + moles_{benz}}$$

The molecular weight of toluene (C_7H_8) is 92.14.

$$\text{moles toluene} = \frac{36.2 \text{ g}}{92.14 \text{ g/mole}} = 0.393 \text{ moles}$$

The molecular weight of benzene (C_6H_6) is 78.11.

$$\text{moles benzene} = \frac{80.6 \text{ g}}{78.11 \text{ g/mole}} = 1.03 \text{ moles}$$

$$X_{tol} = \frac{0.393}{1.03 + 0.393} = 0.276 \qquad X_{benz} = \frac{1.03}{1.03 + 0.393} = 0.724$$

$$P_{tol} = X_{tol} \cdot P^\circ_{tol} \qquad\qquad P_{benz} = X_{benz} \cdot P^\circ_{benz}$$

$$= .276 \,(290 \text{ torr}) \qquad\qquad = .724 \,(745 \text{ torr})$$

$$= 80.0 \text{ torr} \qquad\qquad\qquad = 539 \text{ torr}$$

$$P_{solution} = 80.0 \text{ torr} + 539 \text{ torr} = 619 \text{ torr}$$

Boiling Point Elevation

The boiling point of a liquid is the temperature at which the equilibrium vapor pressure of the liquid equals the pressure of the atmosphere around the liquid. At this temperature, the vapor pressure of the liquid is large enough for the liquid to form bubbles full of vapor.

At sea level, water boils at 100.0°C, when its equilibrium vapor pressure is 760.0 torr. In Denver, at 1.6 km altitude, where the atmospheric pressure is about 640 torr, water boils at 95.5°C.

Raoult's law indicates that adding a non-volatile solute to a solvent lowers the vapor pressure of the solvent at a particular temperature. We will have to heat the solution higher than the boiling point of the pure solvent to raise the vapor pressure of the solution to the atmospheric pressure.

The elevation of the boiling point is proportional to the _molality_ of the solute.

The boiling point elevation:

$$\Delta T_b = T_b - T^\circ_b = K_b \cdot m_{solute}$$

T°_b is the boiling point of the pure solvent.

T_b is the boiling point of the solution.

K_b is characteristic of the solvent, and doesn't depend on the solute.

m is the molality of the solute.

The boiling point of a solution depends only on the concentration of solute _particles_.

Problem 11-6:

What is the boiling point of a solution of 150 grams of sucrose ($C_{12}H_{22}O_{11}$) in 845 grams of water at 1.00 atm?

$$\Delta T_b = K_{b_{H_2O}} \cdot m_{sucrose}$$

$$m_{sucrose} = \frac{moles\ sucrose}{kg\ water}$$

$$moles\ sucrose = \frac{150\ g}{342.2\ g/mole} = 0.438\ moles$$

$$m = \frac{0.438\ moles}{0.845\ kg\ water} = \frac{0.518\ mol}{kg}$$

$$\Delta T_b = \frac{0.52\ deg}{mol/kg} \left(\frac{0.518\ mol}{kg} \right)$$

$$= 0.27°$$

The boiling point is only raised 0.27°! The solution boils at 100.27°C.

Freezing Point Depression

The freezing point of a liquid is the temperature where the vapor pressure of the liquid equals the vapor pressure of its solid. When we add a solute to the liquid, the vapor pressure of the liquid is lowered.

At a lower temperature, the vapor pressure of the pure solid solvent will equal the vapor pressure of the solution. This is the freezing point of the solution. The equation for freezing point depression is similar to the equation for boiling point elevation:

$$T_f^o - T_f = \Delta T_f - K_f \cdot m$$

K_f depends only on the solvent, not on the solute.

The freezing point of the solution depends only on the concentration of solute particles.

Problem 11-7:

What is the freezing point of the sucrose solution in 11-6?

$$\Delta T_f = K_{f_{H_2O}} \cdot m_{sucrose}$$

$$\Delta T_f = \frac{1.86°}{mol/kg} \left(\frac{0.518\ mol}{kg} \right)$$

$$\Delta T_f = 0.963°$$

The freezing point of this sucrose solution is −0.963°C.

Osmotic Pressure

A semi-permeable membrane acts as a fine sieve, allowing small molecules to pass through, stopping the flow of large molecules.

Let's separate a solution containing large solute molecules and solvent from pure solvent by a semi-permeable membrane which allows only solvent molecules to flow through it. Solvent molecules will flow from the pure solvent side of the membrane to the solution side, and from the solution side to the solvent side. The rate of flow of solvent molecules depends on the concentration of the solvent molecules. Solvent molecules will flow to the solution side faster than solvent molecules flow back. There is then a net flow of solvent into the solution, to dilute the solution. This process is osmosis.

We can prevent the flow of solvent across the membrane into the solution by exerting pressure on the solution. The pressure which just prevents the flow of solvent is the osmotic pressure of the solution.

The osmotic pressure is proportional to the concentration of solute.

$$\pi = MRT$$

π is the osmotic pressure in atmospheres

T is the absolute temperature

R is the ideal gas constant, 0.08206 liter atm/mol K

M is the molarity of the solute.

The osmotic pressure depends only on the concentration of the solute particles.

Problem 11-8:

Calculate the osmotic pressure of a sucrose solution of 150.0 grams of sucrose ($C_{12}H_{22}O_{11}$) added to 845 grams of water to form 936 mL of solution at 20°C.

$$\pi = MRT$$

$$\text{Molarity} = \frac{\text{moles sucrose}}{\text{liters solution}}$$

$$\text{moles sucrose} = \frac{150 \text{ g}}{342.2 \text{ g/mole}} = 0.438 \text{ moles}$$

$$m_{sucrose} = \frac{0.438 \text{ moles}}{0.936 \text{ liters}} = 0.468 \text{ } M$$

$$\pi = \frac{0.468 \text{ mol}}{L} \left(\frac{0.08206 \text{ L atm}}{\text{mol K}} \right) (273 + 20 \text{ K})$$

$$= 11.2 \text{ atm!}$$

This pressure can support a mercury column 8.51 meters high!

This is the same sucrose solution we looked at in Problems 11-6 and 11-7.

For this solution:

$$\Delta T_b = 0.27°C$$

$$\Delta T_f = 0.963°C$$

and $\qquad \pi = 11.0 \text{ atm}$

The osmotic pressure of a solution is clearly much more sensitive to the solute concentration than either boiling point elevation or freezing point depression. We can use the osmotic pressure of a solution to determine concentrations of very dilute solutions.

When a semi-permeable membrane separates two different solutions, solvent tends to flow from the solution of lower osmotic pressure to the solution of higher osmotic pressure, that is, solvent flows from the dilute solution into the concentrated solution.

These properties, osmotic pressure, boiling point elevation and freezing point depression are often called "colligative" ("tied together") properties. Each depends only on the amount of solute present. Each can be related to the effect of diluting the solvent concentration by adding solute. This lowers the vapor pressure of the solvent, lowers the freezing point, raises the boiling point and increases the osmotic pressure.

Colligative Properties of Electrolyte Solutions

Freezing point depression, boiling point elevation and osmotic pressure depend on the concentrations of solute particles. When one mole of sodium chloride (NaCl) is dissolved in water, two moles of ions are produced. When one mole of calcium chloride is dissolved, three moles of ions are produced.

We can apply our equations for colligative properties to electrolyte solutions if we include i, the number of moles of ions produced by one mole of solute.

$$\Delta T_b = iK_b m$$

$$\Delta T_f = iK_f m$$

and $\qquad \pi = iMRT$

where i • m is the molality of solute particles in the solution and i • m is the molarity of solute particles in solution.

Problem 11-9:

What do you predict the freezing point of a 0.100 M aqueous solution of $ZnCl_2$ to be?

Solution:

$$\Delta T_f = i \cdot K_f \cdot m$$

$$= i \left(\frac{1.86°C}{mol/kg}\right) \left(\frac{0.100 \ mol}{kg}\right)$$

How many ions are produced by one mole of $ZnCl_2$?

$$ZnCl_2 \rightarrow Zn^{2+} + 2 \ Cl^-$$

Therefore:

$$i = 3:$$

$$\Delta T_f = 3 \times \left(\frac{1.86°C}{mol/kg}\right) \times \left(\frac{0.100 \ mol}{kg}\right)$$

$$= 0.558°C$$

$$T_f = -0.558°C$$

The actual freezing point of this solution ($ZnCl_2$) is found to be −0.495°C. If we use this freezing point to calculate i, we find i = 2.66. i does approach 3 for $ZnCl_2$ in very dilute solutions. We believe that i is less than 3 in more concentrated solutions because the ions are strongly attracted to each other, and don't behave as "independent particles" in concentrated solutions.

Colligative Properties and Molecular Weights

We can use any colligative property to determine the molecular weight of the solute.

Problem 11-10:

When 0.185 g of a protein is dissolved in water to give 25.0 mL of solution the osmotic pressure is 8.23 mm Hg at 25°C. Calculate the molecular weight of the protein.

Solution:

Plan: $\pi \rightarrow$ molarity $\rightarrow$ MW

$$\pi = MRT$$

Convert π in mm Hg to atmospheres:

$$760.00 \text{ mm Hg} = 1 \text{ atm}$$

$$\pi = 8.23 \text{ mm Hg} \times \frac{1 \text{ atm}}{760 \text{ mm Hg}} = 1.08 \times 10^{-2} \text{ atm}$$

$$\pi = MRT$$

$$1.08 \times 10^{-2} \text{ atm} = M \left(\frac{0.08206 \text{ L atm}}{\text{K mol}} \right) (298 \text{ K})$$

Solve for M, the molarity of the protein solution:

$$M = \frac{1.08 \times 10^{-2} \text{ atm}}{\frac{.8206 \text{ L atm}}{\text{K mol}} (298 \text{ K})}$$

$$= \frac{4.42 \times 10^{-4} \text{ mol}}{\text{L}}$$

Therefore, when 0.185 g of this protein is dissolved in water to give 25.0 mL of solution, the molarity is 4.42×10^{-4} moles/liter.

How many moles of protein are in 25.0 mL of solution?

Moles of protein in 25.0 mL of solution = M × volume

$$= \frac{4.42 \times 10^{-4} \text{ moles}}{\text{liters}} \times .025 \text{ L}$$

$$= 1.10 \times 10^{-5} \text{ moles}$$

Then, 0.185 g of protein is 1.10×10^{-5} moles of protein.

The molecular weight of this protein is $\dfrac{0.185 \text{ g}}{1.10 \times 10^{-5}} = \dfrac{1.68 \times 10^4 \text{ g}}{\text{mol}}$

$$= 1.68 \times 10^4 \text{ Daltons}$$

(Biochemists use Daltons as the unit of molecular weight, where 1 Dalton = 1 g/mol.)

11-1. 25.0 g of ethanol (C_2H_5OH) is added to 75.0 g of water to form 96.4 mL of solution.

    a. What is the percent by mass of ethanol?

    b. What is the molarity of ethanol?

    c. What is the molality of ethanol?

    d. What is the mole fraction of ethanol?

    e. What <u>volume</u> of the solution will contain 0.100 moles of ethanol?

11-2. In each pair below, which do you expect to be more soluble in water? Explain your answer.

    a. $CH_3CH_2CH_2CH_2OH$ $HOCH_2CH_2CH_2OH$

    b. $CH_3CH_2OCH_2CH_3$ $CH_3CH_2CH_2CH_2OH$

    c. $CH_3CH_2CH_2CH_2OH$ $CH_3CH_2CH_2OH$

11-3. Use the solubility rules to decide if a precipitate forms when the following solutions are mixed. If a precipitate forms, write the net ionic equation.

    a. $Pb(NO_3)_2(aq)$ and $Na_2SO_4(aq)$

    b. $KCl(aq)$ and $NH_4NO_3(aq)$

    c. $BaCl_2(aq)$ and $AgNO_3(aq)$

    d. $Na_2S(aq)$ and $Zn(NO_3)_2(aq)$

11-4. At 25°C and 1.0 atm pressure of CO_2, the dissolved concentration of CO_2 is 0.034 moles per liter of water. What will the dissolved concentration of CO_2 be when the total pressure of CO_2 is 1.5 atm?

11-5. At 100°C, the equilibrium vapor pressure of pure water is 760 torr. What is the equilibrium vapor pressure of a solution of 0.25 moles of glucose, $C_6H_{12}O_6$, in 90.0 grams of water at 100°C, if the solution obeys Raoult's law.

11-6. The vapor pressure of liquid A at 75°C is 125 torr. The vapor pressure of liquid B at 75°C is 248 torr. If a solution containing 2.00 moles of A and 3.50 moles of B obeys Raoult's law, what is the vapor pressure of the solution at 75°C?

11-7. 0.100 moles of glucose is dissolved in 100.0 g of water.

    a. What is the freezing point of this solution at 1 atm pressure?

b. What is the boiling point of this solution at 1 atm pressure?

$$K_{f(H_2O)} = -1.86°C/m \qquad\qquad K_{b(H_2O)} = 0.52°C/m$$

11-8. 0.100 moles of glucose is dissolved in water to give 125 mL of solution. What is the osmotic pressure of this solution at 25°C?

11-9. 0.100 moles of KNO_3 is dissolved in 100.0 g of water. Calculate the expected freezing point of this solution.

11-10. An aqueous solution containing 2.00 g of a protein in 1.00 liter of water has an osmotic pressure of 4.8 torr at 25°C.

a. What is the molarity of this solution?

b. What is the molecular weight of the protein?

CHAPTER ELEVEN: SELF-TEST ANSWERS

11-1. a. 25.0% by mass

b. 0.543 moles of C_2H_5OH/0.0964 L = 5.63 *M*

c. 0.543 moles of C_2H_5OH/0.0750 kg H_2O = 7.24 m

d. 0.543 moles of C_2H_5OH/(0.543 moles of C_2H_5OH + 4.163 moles H_2O)

= 0.115

e. $0.100 \text{ moles} \times \dfrac{1 \text{ L}}{5.63\ M} = 1.78 \times 10^{-2} \text{ L}$

11-2. a. $HOCH_2CH_2CH_2CH_2OH$ is more soluble

b. $CH_3CH_2CH_2CH_2OH$ is more soluble

c. $CH_3CH_2CH_2OH$ is more soluble

11-3. a. $Pb^{2+} + SO_4^{2-} \rightarrow PbSO_4(s)$

b. No precipitate forms: KNO_3 and NH_4Cl are both soluble.

c. $Ag^+ + Cl^- \rightarrow AgCl(s)$

d. $Zn^{2+} + S^{2-} \rightarrow ZnS(s)$

11-4. 0.051 *M*

11-5. $P_{H_2O_{soln}} = X_{H_2O} \cdot P_{H_2O_{pure}}$

$= (0.952)(760 \text{ torr}) = 724 \text{ torr}$

11-6. $P_{soln} = X_A P_A^\circ + X_B P_B^\circ = (0.364)(125) + (0.636)(248)$

$= 203 \text{ torr}$

11-7. The solution is 1.00 <u>molal</u>.

a. − 1.86°C

b. 100.52°C

11-8. $\pi = MRT;\ M = \dfrac{0.100 \text{ moles}}{0.125 \text{ L}} = 0.800\ M$

$\pi = 19.6 \text{ atm}$

11-9. a. −3.72°C

11-10. a. $M = \dfrac{\pi}{RT}$

$$= \left(\dfrac{4.8\ \text{torr}}{760\ \text{torr/atm}}\right) \dfrac{1}{\left(\dfrac{0.08206\ \text{L atm}}{\text{K mol}}\right)(298\ \text{K})}$$

$= 2.6 \times 10^{-4}\ M$

b. $2.00\ \text{g} = 2.6 \times 10^{-4}\ \text{moles}$

$\text{MW} = 7.7 \times 10^{3}\ \text{g/mole}$

CHAPTER TWELVE – CHEMICAL KINETICS

Chemical kinetics is the study of rates of reactions, and how the rate changes as conditions are varied. Why study kinetics? Kinetics can give us insight into what might be happening at the molecular level.

Reaction Rate

The rate of reaction is always expressed as the change in concentration (of a reactant or product) per unit time. Consider a general reaction:

$$A + 2 B = 2 C + D$$

(In this chapter we will use = to indicate an overall reaction.)

We could describe the rate in terms of the <u>appearance</u> of <u>D</u>.

$$\text{rate} = \frac{[D]_{time_2} - [D]_{time_1}}{time_2 - time_1} = \frac{\Delta[D]}{\Delta t}$$

where [D] is the concentration of D in moles/liter, or in terms of the <u>disappearance</u> of <u>A</u>:

$$\text{rate} = -\left(\frac{[A]_{time_2} - [A]_{time_1}}{time_2 - time_1}\right) = -\frac{\Delta[A]}{\Delta t}$$

Rates of reactions are always positive. Any rate involving the disappearance of a reactant will include a negative sign.

Problem 12-1:

a. How does the rate of appearance of <u>C</u> compare to the rate of appearance of <u>D</u>?

Solution:

The stoichiometry of the reaction is:

$$A + 2 B = 2 C + D$$

For every mole of <u>D</u> formed, <u>2</u> moles of <u>C</u> are formed.

Then:
$$\Delta[C] = 2 \Delta[D]$$

$$\frac{\Delta[C]}{\Delta t} = \frac{2 \Delta[D]}{\Delta t}$$

b. How does the rate of disappearance of <u>A</u> compare to the rate of appearance of <u>D</u>?

Solution:

For every mole of <u>D formed</u>, one mole of <u>A reacts</u>:

[D] increases, as [A] decreases

$$\Delta[D] = -\Delta[A]$$

$$\frac{\Delta[D]}{\Delta t} = \frac{-\Delta[A]}{\Delta t}$$

Rates of reactions vary with concentrations of reactants, concentrations of products, and with temperature. First, let's look at the effect of concentrations.

Following Reactions

How can we study the rate of a reaction? We need to measure the concentrations of the reactants at various times while the reaction continues. If the reaction is slow enough, we can remove a sample of the reaction mixture and titrate the sample to determine concentrations directly. Or, we can measure some property of the solution which is proportional to the concentration of a reactant or product. If we carry out this reaction:

$$2 \, NO_2(g) = 2 \, NO(g) + O_2(g)$$

and determine the concentrations indirectly in a "constant-volume" container, the pressure of the gas mixture increases as products form. If we know the volume of the container, we can calculate the moles of gas present at any time from the total pressure.

For some reactions we can follow the formation or disappearance of colored substances like Br_2 (aq) or MnO_4^- (aq) by measuring the intensity of the color of a solution. We could study the reaction of bromine with acetone:

$$CH_3\overset{\overset{\displaystyle O}{\|}}{C}CH_3 + Br_2 = CH_3\overset{\overset{\displaystyle O}{\|}}{C}CH_2Br + H^+ + Br^-$$

by following the disappearance of the color due to bromine. We would need to "calibrate" the intensity of the color of the solution with a solution of known Br_2 concentration.

Determining Rates of Reactions

The rate of a reaction is the change in concentration (of a reactant or product) per unit time.

$$rate = \frac{\Delta \; concentration}{\Delta \; time}$$

If we plot a graph of concentration versus time (concentration on the y-axis; time on the x-axis), the rate of the reaction at any time is the magnitude of the slope of the curve at that point.

In Figure 2.2, the slope of the curve changes with time. The rate of the reaction is not constant. Look at the data in Table 12.3:

Over the first 2000 seconds, the average rate of the reaction is:

$$\text{rate} = -\frac{\Delta[N_2O_5]}{\Delta t} = -\frac{(0.88 \text{ moles/L} - 1.00 \text{ moles/L})}{200 \text{ s} - 0 \text{ s}}$$

$$= 6.0 \times 10^{-4} \text{ moles/L/s}$$

During the time from 600 s to 800 s, the average rate is

$$\text{rate} = -\frac{\Delta[N_2O_5]}{\Delta t} = -\frac{(0.61 \text{ moles/L} - 0.69 \text{ moles/L})}{800 \text{ s} - 600 \text{ s}}$$

$$= 4.0 \times 10^{-4} \text{ moles/L/s}$$

As the concentration of N_2O_5 drops, the rate of the reaction decreases.

Rate Law

The rate law for a reaction gives the relationship between the rate of the reaction and the concentrations of the reactants (and products). The rate law must be determined by experiment. We cannot predict the rate law from the overall reaction.

For the reaction

$$2 \text{ N}_2O_5(\text{soln}) = 4 \text{ NO}_2(\text{soln}) + O_2(g) \text{ at } 45°C.$$

When:
$$[N_2O_5] = 0.90 \text{ } M, \text{ rate} = 5.4 \times 10^{-4} \text{ moles/L/s}$$

$$[N_2O_5] = 0.60 \text{ } M, \text{ rate} = 3.6 \times 10^{-4} \text{ moles/L/s}$$

$$[N_2O_5] = 0.30 \text{ } M, \text{ rate} = 1.8 \times 10^{-4} \text{ moles/L/s}$$

The general form of the rate law is:

$$\text{rate} = k[N_2O_5]^n$$

How do we determine n?

Compare two reaction mixtures:

$$\text{rate}_2 = k[N_2O_5]_2{}^n$$
$$\text{rate}_1 = k[N_2O_5]_1{}^n$$

$$\frac{\text{rate}_2}{\text{rate}_1} = \left(\frac{[N_2O_5]_2}{[N_2O_5]_1}\right)^n$$

$$\frac{5.4 \times 10^{-4} \text{ moles/L/s}}{3.6 \times 10^{-4} \text{ moles/L/s}} = \left(\frac{.90 \text{ } M}{.60 \text{ } M}\right)^n$$

$$1.5 = (1.5)^n$$

This reaction is first-order:

$$\text{rate} = k[N_2O_5]^1$$

Problem 12-2:

What is the rate constant (including units) for the decomposition of N_2O_5 in carbon tetrachloride at 45°C.

Solution:

$$\text{rate} = k[N_2O_5]^1$$

Solve for k:

$$k = \frac{\text{rate}}{[N_2O_5]}$$

When $[N_2O_5] = 0.90$ M: rate = 5.4×10^{-4} moles/L/s

$$k = \frac{5.4 \times 10^{-4} \text{ moles/L/s}}{0.90 \text{ moles/L}} = 6.0 \times 10^{-4}/s$$

$$= 6.0 \times 10^{-4} \text{ s}^{-1}$$

Check: when $[N_2O_5] = 0.60$ M: rate = 3.6×10^{-4} moles/L/s

$$k = \frac{3.6 \times 10^{-4} \text{ moles/L/s}}{0.60 \text{ moles/L}} = 6.0 \times 10^{-4} \text{ s}^{-1}$$

If we have the right rate law, k should be constant (within experimental error).

How can we determine the rate law for more complicated reactions? We can carry out a series of experiments, varying the initial concentrations of the reactants. Compare two experiments where the concentrations of all but one reactant are the same.

Problem 12-3

   Use the data to determine the rate law for the reaction

$$A + B + C = 2 D + E \text{ at } 25°C$$

Experiment	[A]	[B]	[C]	initial rate $= -\dfrac{\Delta[B]}{\Delta t}$
1	.200 M	.150	.300	8.00×10^{-4} mol/L/s
2	.400	.150	.300	8.00×10^{-4} mol/L/s
3	.200	.150	.900	24.0×10^{-4} mol/L/s
4	.400	.225	.300	18.0×10^{-4} mol/L/s

Solution:

   The rate law has the general form:

$$\text{rate} = k[A]^a [B]^b [C]^c$$

   To determine a, we can compare experiments 1 and 2: the concentration of B is the same in both experiments, the concentration of C is the same in both experiments, but the concentrations of A differ:

$$\frac{\text{rate}_2}{\text{rate}_1} = \frac{k}{k} \cdot \frac{[A]_2^a}{[A]_1^a} \cdot \frac{[B]_2^b}{[B]_1^b} \cdot \frac{[C]_2^c}{[C_1]^c}$$

$$\frac{\text{rate}_2}{\text{rate}_1} = \frac{8.00 \times 10^{-4} \text{ mol/L/s}}{8.00 \times 10^{-4} \text{ mol/L/s}} = \frac{[.400\ M]^a}{[.200\ M]^a}$$

$$1 = 2^a$$

$$a = 0: \quad 1 = 2^0$$

   That is, the rate doesn't depend on the concentration of A.

To determine b:

   Compare two experiments which differ only in the concentration of B. Compare experiments 2 and 4.

$$\frac{\text{rate}_4}{\text{rate}_2} = \frac{k}{k} \cdot \frac{[A]_4^a}{[A]_2^a} \cdot \frac{[B]_4^b}{[B]_2^b} \cdot \frac{[C]_4^c}{[C]_2^c}$$

$$\frac{18.0 \times 10^{-4} \text{ mol/L/s}}{8.0 \times 10^{-4} \text{ mol/L/s}} = \left(\frac{.225\ M}{.150\ M}\right)^b$$

$$2.25 = (1.50)^b$$

$$b = 2$$

To determine c:

Compare experiments 1 and 3.

$$\frac{rate_3}{rate_1} = \frac{[C]_3^c}{[C]_1^c}$$

$$\frac{24.0 \times 10^{-4} \ mol/L/s}{8.0 \times 10^{-4} \ mol/L/s} = \left(\frac{0.900 \ M}{.300 \ M}\right)^c$$

$$3 = (3)^c$$

$$c = 1$$

The rate law is:

$$rate = k[A]^0[B]^2[C]^1$$

$$= k[B]^2[C]^1$$

Problem 12-4:

What is the rate constant at 25°C for the reaction in Problem 12-3?

$$A + B + C = 2 \ D + E$$

Solution:

The rate law is:

$$rate = k[B]^2[C]$$

Solve for k:

$$k = \frac{rate}{[B]^2[C]}$$

We can use any experiment: From experiment 1:

$$k = \frac{8.00 \times 10^{-4} \ mol/L/s}{[0.150 \ M]^2[0.300 \ M]}$$

$$= 0.118 \ M^{-2}s^{-1}$$

From experiment 4:

$$k = \frac{18.00 \times 10^{-4} \ mol/L/s}{[0.225 \ M]^2[0.300 \ M]}$$

$$= 0.118 \ M^{-2}s^{-1}$$

The Integrated Rate Law

The rate laws we have used so far are "differential" rate laws, rate laws involving "differences," $\Delta[A]$ and Δt. These give the rate as a function of concentration. The "integrated" rate law is a new form of the rate law which gives the reactant concentration(s) as a function of time.

First Order

If the "differential" rate law is:

$$rate = -\frac{\Delta[A]}{\Delta t} = k[A],$$

The integrated form of the rate law is:

$$[A] = [A]_0 e^{-kt}$$

or, taking the natural logarithm of both sides:

$$\ln[A] = \ln[A]_0 - kt$$

The concentration of A decreases exponentially with time.

where

$\quad$ [A] is the concentration of A at time t

$\quad$ $[A]_0$ is the initial concentration of A

$\quad$ k is the first order rate constant

A plot of $\ln[A]$ versus time will yield a straight line with slope $-k$.

Problem 12-5:

For the reaction

$$2\ N_2O_5(\text{soln}) = 4\ NO_2(\text{soln}) + O_2(g) \text{ at } 45°$$

The rate law is: $\text{Rate} = 6.0 \times 10^{-4}\ s^{-1}\ [N_2O_5]$

If the initial concentration of N_2O_5 is 1.20 M, what is the concentration after 300 seconds?

$$\ln[N_2O_5] = \ln[N_2O_5]_0 - kt$$

$$= \ln(1.20) - 6.0 \times 10^{-4}\ s^{-1}\ [300\ s]$$

$$= \ln(1.20) - 0.180 = .0023$$

$$[N_2O_5] = e^{.0023} = 1.00\ M$$

Second Order Reactions

The "differential" rate law for a second order reaction is:

$$rate = k[A]^2$$

The integrated rate law for this reaction is:

$$\frac{1}{[A]} = kt + \frac{1}{[A]_0}$$

Problem 12-6:

For the dimerization (forming dimers) of butadiene:

$$2 \ C_4H_6(g) = C_8H_{12}(g)$$

The rate law is found to be:

$$rate = k[C_4H_6]^2$$

$$with \ k = \frac{6.14 \times 10^{-2} \ L}{mol \ s}$$

If the initial concentration of C_4H_6 is 0.0150 moles/liter, what will the concentration be after 2000 seconds?

Solution:

$$\frac{1}{[C_4H_6]} = \frac{6.14 \times 10^{-2} \ L}{mol \ s} \ (2000 \ s) + \frac{1}{0.0150 \ moles/L}$$

$$\frac{1}{[C_4H_6]} = \frac{122._8 \ L}{mol} + \frac{66.7 \ L}{mol} = \frac{189 \ L}{mol}$$

$$[C_4H_6] = 0.00528 \ moles/L$$

Zero-order Reactions

The differential rate law for a zero-order reaction is

$$rate = k[A]^0 = k$$

The rate of a zero-order reaction is constant.

The integrated rate law for a zero-order reaction is:

$$[A]_0 - [A] = kt$$

$$or \ \ [A] = [A]_0 - kt$$

Half-life

The half-life for a reaction is the time required for a reactant to reach half of its concentration.

For <u>first-order reactions</u>, the rate law is: Rate = k[A] and the half-life is:

$$t_{1/2} = \frac{0.693}{k}$$

The half-life for a first-order reaction is constant.

For <u>second-order reactions</u> the rate law is: rate = k[A]^2 and the half-life is:

$$t_{1/2} = \frac{1}{k[A]_0}$$

The half-life for a second-order reaction depends on the concentration. During the first half-life:

$$t_{1/2} = \frac{1}{k[A]_0}$$

[A]$_0$ decreases by 1/2.

At the beginning of the second half-life, the concentration of A is [A]$_0$/2. Then, $t_{1/2}$ for the second half-life is:

$$t_{1/2} = \frac{1}{k\left(\frac{[A]_0}{2}\right)} = \frac{2}{k[A]_0}$$

For second-order reactions, each successive $t_{1/2}$ is <u>double</u> the previous one.

For <u>zero-order reactions</u> the rate law is: Rate = k[A]^0 = k.

$$t_{1/2} = \frac{[A]_0}{2k}$$

For zero-order reactions, each $t_{1/2}$ is <u>half</u> the previous one.

If we compare the length of the first half-life of a reaction to the length of the second we should be able to see whether the reaction is zero, first, or second order.

Table 12.6 in the text summarizes the characteristics of zero, first, and second order rate laws involving only one reactant.

For more complex reactions, we can use reaction mixtures with large concentrations of all but one reactant. Then only one reactant concentration varies appreciably with time. The other reactant concentrations are essentially constant.

If we study the reaction:

$$6 \; H^+ + BrO_3^- + 5 \; Br^- = 3 \; H_2O + 3 \; Br_2$$

The general form of the rate law is:

$$\frac{\Delta[BrO_3^-]}{\Delta t} = -k[H^+]^a[BrO_3^-]^b[Br^-]^c$$

In mixtures with hydrogen and bromide ion $[Br^-]$ in large excess, we observe first order kinetics for the disappearance of BrO_3^- in each mixture:

$$\frac{\Delta[BrO_3^-]}{\Delta t} = -k_{obs}[BrO_3^-]$$

Where the observed first-order rate constant, k_{obs}, depends on the hydrogen ion and bromide ion concentrations:

$$k_{obs} = k[H^+]^a[Br^-]^c$$

Experiment	k_{obs}	$[H^+]_0$	$[Br^-]_0$
1	$0.90 \times 10^{-2} \; s^{-1}$	$0.100 \; M$	$0.100 \; M$
2	$1.80 \times 10^{-2} \; s^{-1}$	$0.100 \; M$	$0.200 \; M$
3	$3.60 \times 10^{-2} \; s^{-1}$	$0.200 \; M$	$0.100 \; M$

Compare k_{obs} in Experiments 1 and 2, where the initial bromide ion concentrations differ.

$$\frac{k_{obs\,2}}{k_{obs\,1}} = \frac{k[H^+]_2^a[Br^-]_2^c}{k[H^+]_1^a[Br^-]_1^c}$$

$$\frac{1.80 \times 10^{-2} \; s^{-1}}{0.90 \times 10^{-2} \; s^{-1}} = \left(\frac{.200 \; M}{.100 \; M}\right)^c$$

$$2 = 2^c$$

Therefore $c = 1$

Compare experiments 3 and 1: We find $a = 2$. Therefore, for this reaction, the overall rate law is:

$$rate = -\frac{\Delta[BrO_3^-]}{\Delta t} = k[BrO_3^-][H^+]^2[Br^-]^1$$

Reaction Mechanisms

The balanced equation for a reaction tells us the overall stoichiometry for the reaction but does not tell us how the reaction takes place. Most chemical reactions take place as a series of <u>elementary</u> reactions. Each elementary step involves molecules or ions that <u>collide.</u> Some of these collisions involve rearrangements of atoms to produce new molecules or ions. The rate of an elementary step is proportional to the number of collisions between the reacting molecules or ions.

The mechanism of a reaction is a sequence of elementary steps, the pathway for the reaction. We can never <u>prove</u> that a mechanism is "true," but we can show that a mechanism is consistent with our experimental observations.

From Mechanism to Rate Law

If we have a series of elementary steps in a mechanism, the overall rate of the reaction will equal the rate of the slowest step, the "rate-determining-step." The rate-determining-step is a bottle-neck.

For the reaction between $NO_2(g)$ and $CO_2(g)$ [$NO_2(g) + CO(g) = NO(g) + CO_2(g)$] the proposed mechanism has two steps:

Mechanism I:

$$NO_2(g) + NO_2(g) \xrightarrow{k_1} NO_3(g) + NO(g) \quad \text{Slow}$$

$$NO_3(g) + CO(g) \xrightarrow{k_2} CO_2(g) + NO_2(g) \quad \text{Fast}$$

$NO_3(g)$ is an <u>intermediate</u>: $NO_3(g)$ does not appear in the overall reaction. $NO_3(g)$ is <u>produced and</u> consumed in the overall process.

The first step is the rate-determining-step. $NO_3(g)$ reacts with $CO(g)$ as soon as it forms. The rate of the overall reaction is the rate of the first step. The rate of the first step is proportional to the number of collisions between $NO_2(g)$ molecules.

$$\text{rate} = k_1[NO_2]^2$$

This rate law agrees with the experimentally observed rate law. The concentration of $CO(g)$ does not influence the rate of the reaction.

Suppose Step 1 had been fast compared to Step 2: what rate law would we expect? If the forward rate of Step 1 were fast, concentrations of NO_3 and NO would increase. Then NO_3 could either react with NO to produce NO_2, or NO_3 could react with CO to produce products.

Mechanism II:

$$NO_2(g) + NO_2(g) \underset{k_{-1}}{\overset{k_1}{\rightleftharpoons}} NO_3(g) + NO(g) \qquad \text{Fast}$$

$$NO_3(g) + CO(g) \xrightarrow{k_2} CO_2(g) + NO(g) \qquad \text{Slow}$$

The rate of the reaction would be the rate of the slow step:

$$\text{rate} = k_2[NO_3][CO]$$

NO_3 is an intermediate; we don't know its concentration. Can we determine the concentration of NO_3 in terms of the reactants and products? In the simplest case, Step 1 and the reverse of Step 1 are both fast compared to Step 2. Then, there is a balance between the forward rate of Step 1 and the backward rate:

$$k_1[NO_2]^2 = k_{-1}[NO_3][NO]$$

and
$$[NO_3] = \frac{k_1[NO_2]^2}{k_{-1}[NO]}$$

Then the rate for the overall reaction would be:

$$\text{rate} = k_2[NO_3][CO] = \frac{k_2 k_1}{k_{-1}} \frac{[NO_2]^2[CO]}{[NO]}$$

This is not the observed rate law. The first mechanism, with the first step rate-determining, does give the observed rate law. We cannot prove that the first mechanism is "right"; we have proved that the second mechanism is not "right."

The Collision Model of Kinetics

The rate of an elementary step is proportional to the number of collisions between the reacting species. However, the rate of reaction is much smaller than the calculated rate of collisions. Why doesn't every collision lead to products?

In every reaction, some bonds are broken, while new bonds form. The energy to begin breaking the bonds must come from the kinetic energies of the colliding particles. There is a potential energy "barrier" to the reaction. At any given temperature, only a fraction of the collisions will have enough energy to overcome the barrier. At a higher temperature, a greater fraction of the collisions will have enough energy.

Even though a particular collision has enough energy to overcome the barrier, the orientation of the colliding molecules may not lead to reaction. Figure 12-11 gives examples of collisions between BrNO molecules in different orientations. Only some of the orientations can lead to Br_2 and NO.

The Arrhenius equation collects all these factors into an expression for the rate constant:

$$k = pze^{\frac{-E_a}{RT}} = Ae^{\frac{-E_a}{RT}}$$

Where: z is the frequency of collisions

p is the fraction of collisions with an effective orientation

$e^{-E_a/RT}$ is the fraction of collisions with energy greater than E_a

E_a is the potential energy barrier: the energy of activation for the reaction.

Taking the natural logarithm of both sides of the Arrhenius equation:

$$\ln k = -\frac{E_a}{R}\left(\frac{1}{T}\right) + \ln A$$

If we measure the rate constant of a reaction at different temperatures, we can determine the energy of activation by plotting ln k versus 1/T. The slope of this line is $-E_a/R$.

From Rate Law to Mechanism

A mechanism, the series of elementary steps <u>and</u> their relative rates, must give the correct overall reaction and predict the observed rate law. Once we determine the rate law, how can we develop a mechanism?

Step 1:

Look at the rate law: use the rate law to determine the composition of the activated complex, the collection of atoms at the top of the energy barrier, for the rate-determining step: add together the species in the numerator, subtract the species in the denominator.

a. For example: For the reaction:

$$NO_2(g) + CO(g) = NO(g) + CO_2(g)$$

The rate law is: Rate $= k\,[NO_2]^2$

The formula for the activated complex for the rate-determining step is $NO_2 + NO_2 = (N_2O_4)$.

b. For the reaction:

$$2\,O_3(g) \rightarrow 3\,O_2(g)$$

The observed rate law is: Rate $= k[O_3]^2/[O_2]$

The formula for the activated complex for the rate-determining step is $O_3 + O_3 - O_2 = (O_4)$.

c. For the gas phase reaction of chlorine with chloroform:

$$Cl_2(g) + CHCl_3(g) = HCl(g) + CCl_4(g)$$

The rate law is: Rate $= k[Cl_2]^{1/2}[CHCl_3]$

The formula for the activated complex for the rate-determining step is: $1/2(Cl_2) + CHCl_3 = (CHCl_4)$

Step 2

Write a possible reaction for the rate-determining step, using reactants and intermediates, producing products.

a. If the activated complex is (N_2O_4): $NO_2 + NO_2 \rightarrow NO_3 + NO$

b. If the activated complex is (O_4): $O_3 + O \rightarrow 2 O_2$

c. If the activated complex is $(CHCl_4)$: $Cl + CHCl_3 \rightarrow CCl_3 + HCl$

Step 3

If intermediates appear in the rate-determining step, write reactions preceding that step that produce those intermediates: these must be fast compared to the rate determining step.

b. $O_3 \rightleftharpoons O_2 + O$

c. $Cl_2 \rightleftharpoons 2 Cl$

Step 4

Write reactions which follow the rate-determining step to give the over-all stoichiometry.

a. $NO_2 + NO_2 \rightarrow NO_3 + NO$ Slow

$NO_3 + CO \rightarrow NO_2 + CO_2$ Fast

$NO_2 + NO_2 + NO_3 + CO = NO_3 + NO + NO_2 + CO_2$ Overall

b. $O_3 \rightleftharpoons O_2 + O$ Fast

$O + O_3 \rightarrow 2 O_2$ Slow

$O_3 + O + O_3 = O_2 + O + 2 O_2$ Overall

c. $Cl_2 \rightleftharpoons 2\ Cl$ Fast

$Cl + CHCl_3 \rightarrow CCl_3 + HCl$ Slow

$CCl_3 + Cl \rightarrow CCl_4$ Fast

$Cl + Cl + Cl_2 + CHCl_3 + CCl_3 = CCl_3 + HCl + 2\ Cl + CCl_4$ Overall

Catalysts

A catalyst provides another path for a reaction, a path with a lower activation energy than the uncatalyzed path. The catalyst increases the reaction rate but is not consumed by the reaction. The catalyst does not appear in the overall reaction, but does appear in the rate law.

Bromine can catalyze the decomposition of hydrogen peroxide in solution. The overall reaction is:

$$2\ H_2O_2(aq) = 2\ H_2O(\ell) + O_2(g)$$

The mechanism of the catalyzed reaction is thought to be:

$$Br_2 + H_2O_2 \rightarrow 2\ Br^- + 2\ H^+ + O_2(g)$$

$$2\ Br^- + H_2O_2 + 2\ H^+ \rightarrow Br_2 + 2\ H_2O$$

Manganous ion can catalyze the reaction between cerium (IV) ions and thalium (I) ions in aqueous solution. The overall reaction is:

$$2\ Ce^{+4}(aq) + Tl^+(aq) \rightarrow 2\ Ce^{+3}(aq) + Tl^{+3}(aq)$$

The mechanism of the catalyzed reaction is thought to be:

$$Mn^{+2} + Ce^{+4} \rightarrow Ce^{+3} + Mn^{+3}\quad \text{Slow}$$

$$Mn^{+3} + Ce^{+4} \rightarrow Ce^{+3} + Mn^{+4}\quad \text{Fast}$$

$$Mn^{+4} + Tl^+ \rightarrow Tl^{+3} + Mn^{+2}\quad \text{Fast}$$

With the rate law: Rate $= k[Mn^{+2}][Ce^{+4}]$

Manganous ion provides a faster path for the reaction than the uncatalyzed path.

CHAPTER TWELVE: SELF–TEST

12-1.   Given the following data for the reaction A + B → 2 C + D

Experiment	Initial [A]	Concentrations [B]	Initial Rate of Appearance of C
i)	0.0400 M	0.0100 M	2.00 × 10^{-3} $\frac{moles}{liter\ s}$
ii)	0.0600 M	0.0200 M	9.00 × 10^{-3} $\frac{moles}{liter\ s}$
iii)	0.0600 M	0.0100 M	4.50 × 10^{-3} $\frac{moles}{liter\ s}$

a.   Write the rate law for the reaction.

b.   Calculate the rate constant for the reaction.

c.   Calculate the rate of appearance of C when [A] = 0.050 M and [B] = 0.030 M.

d.   In experiment ii, what is the rate of <u>disappearance</u> of B?

12-2.   Compound C decomposes to give products. The rate law, for the decomposition is: Rate = k[C].   At 25°C, k is found to be 3.6 × 10^{-3} s^{-1}.

If the initial concentration of C is 2.0 × 10^{-2} M:

a.   What is the half–life of the decomposition of C?

b.   What is the concentration of C after 380 seconds?

c.   What is the concentration of C after 500 seconds?

12-3.   Why do rates of reactions increase with increasing temperature?

12-4.   Three mechanisms are proposed for the reaction between nitrogen dioxide and ozone: 2 NO_2(g) + O_3(g) → N_2O_5(g) + O_2(g)

i) O_3 + O_3 → O_4 + O_2 slow

O_4 + NO_2 → NO_3 + O_3 fast

NO_3 + NO_2 → N_2O_5 fast

ii) NO_2 + O_3 → NO_3 + O_2 slow

NO_3 + NO_2 → N_2O_5 fast

iii) $NO_2 + NO_2 \rightleftharpoons N_2O_4$ fast

     $N_2O_4 + O_3 \rightarrow N_2O_5 + O_2$ slow

a. Write the expected rate law for each mechanism.

b. If the experimentally determined rate law is:

$$Rate = k[NO_2][O_3],$$

which mechanism could be correct?

12-5. a. Explain how a catalyst affects a reaction rate.

b. Can a catalyst catalyze the forward rate of a reaction while not affecting the reverse rate of the reaction?

CHAPTER TWELVE: SELF-TEST ANSWERS

12-1. a. Rate = $k[A]^2[B]$

b. k = 125

c. Rate = $9.4 \times 10^{-3}\ \frac{\text{moles}}{\text{liter s}}$

d. $\frac{\Delta B}{\Delta t} = \frac{1}{2}\frac{\Delta C}{\Delta t} = -4.50 \times 10^{-3}\ \frac{\text{moles}}{\text{liter s}}$

12-2. a. $t_{1/2} = \frac{0.693}{k} = 1.9 \times 10^2$ s

b. after 2 half-lives: $[C] = 1/4(2.0 \times 10^{-2}\ M) = 5.0 \times 10^{-3}\ M$.

c. $[C] = [C]_o e^{-kt}$

$= (2.0 \times 10^{-2}\ M)e^{-(3.6 \times 10^{-3}\ s^{-1})(500\ s)}$

$= 2.0 \times 10^{-2}\ M)e^{-1.8} = 3.3 \times 10^{-3}\ M$

12-3. At higher temperatures, molecules will collide more frequently, and more collisions will have enough energy to overcome the potential energy barrier.

12-4. a. i) Rate = $k[O_3]^2$

ii) Rate = $[NO_2][O_3]$

iii) Rate = $k_2[N_2O_4][O_3] = \frac{k_1k_2}{k_{-1}}[NO_2]^2[O_3]$

b. Mechanism ii could be correct, because it matches the observed rate law.

12-5. a. A catalyst provides another path for a reaction, a path with a lower activation energy.

b. The catalyzed path has a lower activation energy for both the forward reaction and the reverse reaction.

CHAPTER THIRTEEN   CHEMICAL EQUILIBRIA

In Chapter 10, we looked at the equilibrium vapor pressure above a liquid as a dynamic process: molecules of liquid escape from the liquid into the vapor at the same rate that vapor molecules return to the liquid.

Nitrogen dioxide is a brown gas which dimerizes to form colorless N_2O_4.

$$2\ NO_2(g) \rightleftharpoons N_2O_4(g)$$

When NO_2 is placed in a sealed glass vessel, the brown color gradually decreases as NO_2 is converted to N_2O_4. However, the color does not fade completely. When colorless N_2O_4 is placed in a sealed glass vessel, the reaction mixture becomes brown as NO_2 forms. When the color stops changing the rate of the forward reaction, forming N_2O_4, equals the rate of the reverse reaction, N_2O_4 dissociating to form NO_2. Then the system has reached chemical equilibrium. The equilibrium position of the mixture can be approached from either direction.

For some reactions, the equilibrium favors nearly complete formation of products; in others, the equilibrium favors reactants.

The Equilibrium Constant

When is a system at equilibrium? We find that for any reaction, there is an equilibrium expression which describes the composition of all the equilibrium mixtures. For the general reaction: $jA + kB \rightleftharpoons nC + mD$

The equilibrium expression is: $$K = \frac{[C]^n[D]^m}{[A]^j[B]^k}$$

The equilibrium constant, K, equals the product of the equilibrium concentrations of the products, each raised to the power of its coefficient in the balanced equation divided by the product of the equilibrium concentrations of the reactants, each raised to the power of its coefficient in the balanced equation. (Whew! easier done than said.)

Problem 13-1:

Write the equilibrium expression for each of these reactions

a. $2\ SO_2(g) + O_2(g) \rightleftharpoons 2\ SO_3(g)$

b. $2\ NO_2(g) \rightleftharpoons N_2O_4(g)$

c. $H_2(g) + I_2(g) \rightleftharpoons 2\ HI(g)$

Solution:

a. $K = \dfrac{[SO_3]^2}{[SO_2]^2[O_2]}$

b. $K = \dfrac{[N_2O_4]}{[NO_2]^2}$

c. $K = \dfrac{[HI]^2}{[H_2][I_2]}$

Problem 13-2:

An equilibrium mixture containing SO_3, SO_2 and O_2 at 600°C has the following equilibrium concentrations:

$$[SO_3] = 0.450 \ M$$

$$[SO_2] = 0.225 \ M$$

$$[O_2] = 0.922 \ M$$

Calculate the equilibrium constant for the reactions

a. $2 \ SO_2(g) + O_2(g) \rightleftharpoons 2 \ SO_3(g)$

and b. $2 \ SO_3(g) \rightleftharpoons 2 \ SO_2(g) + O_2(g)$

Solution:

a. $2 \ SO_2(g) + O_2(g) \rightleftharpoons 2 \ SO_3(g)$ 　 b. $2 \ SO_3(g) \rightleftharpoons O_2(g) + 2 \ SO_2(g)$

$$K = \dfrac{[SO_3]^2}{[SO_2]^2[O_2]} \qquad\qquad K = \dfrac{[O_2][SO_2]^2}{[SO_3]^2}$$

$$= \dfrac{[0.450]^2}{[0.225]^2[0.922]} \qquad\qquad = \dfrac{[0.922][0.225]^2}{[0.450]^2}$$

$$= 4.34 \qquad\qquad\qquad = 0.230$$

Notice that the equilibrium constant for the reverse reaction is the inverse of the equilibrium constant for the forward reaction.

$$K_{eq_{reverse}} = \dfrac{1}{K_{eq_{forward}}}$$

Equilibrium constants can be expressed in terms of pressures. Using the ideal gas law: $PV = nRT$, the concentration of a gas in moles/liter, $n/V = P/RT$.

For the reaction $2 \ SO_2(g) + O_2(g) \rightleftharpoons 2 \ SO_3(g)$

$$K = \dfrac{[SO_3]^2}{[SO_2]^2[O_2]} = \dfrac{(P_{SO_3}/RT)^2}{(P_{SO_2}/RT)^2(P_{O_2}/RT)}$$

$$K = \dfrac{P_{SO_3}^2}{P_{SO_2}^2 P_{O_2}} \cdot RT$$

We can define K_p, the equilibrium constant in terms of pressures. For this reaction: $K = K_p \cdot RT$

and
$$K_p = K(RT)^{-1}$$

In general, $K_p = K(RT)^{\Delta n}$ where Δn is the difference between the number of moles of gaseous products and the number of moles of gaseous reactants. For the reaction:

$$2\ SO_2(g) + O_2(g) \rightleftharpoons 2\ SO_3(g).$$

there are 2 moles of gaseous product and 3 moles of gaseous reactants. $\Delta n = 2 - 3 = -1$ and $K_p = K(RT)^{-1}$. Actually, you don't need to memorize this relationship. You can always derive K_p from K using $n/V = P/RT$.

Problem 13-3:

At a certain temperature 0.250 moles of N_2 and 0.800 moles of H_2 are placed in a 2.00 liter flask. At equilibrium 0.100 moles of NH_3 have formed? What is the equilibrium constant for the reaction:

$$N_2(g) + 3\ H_2(g) \rightleftharpoons 2\ NH_3(g)$$

at this temperature?

Solution:

a. Make a table:

	N_2	H_2	NH_3
Initial moles	0.250	0.800	-0-
Reacted	--	--	
Final moles	--	--	0.100
Equilibrium Concentration	--	--	--

b. If 0.100 moles of NH_3 are formed, how many moles of N_2 reacted? 0.050 How many moles of H_2 reacted? 0.150

$$\text{moles } N_2 = 0.100 \text{ moles } NH_3 \times \frac{1 \text{ mole } N_2}{2 \text{ moles } NH_3} = 0.050$$

$$\text{moles } H_2 = 0.100 \text{ moles } NH_3 \times \frac{3 \text{ moles } H_2}{2 \text{ moles } NH_3} = 0.150$$

c. Complete the table

	N_2	H_2	NH_3
Initial moles	0.250	0.800	–0–
Reacted	0.050	0.150	
Final moles	0.200	0.650	0.100
Equilibrium Concentration	0.100 M	0.325 M	0.050 M

d. Calculate the equilibrium constant:

$$K = \frac{[NH_3]^2}{[N_2][H_3]^3} = \frac{(0.050)^2}{(0.100)(0.325)^3} = 0.73$$

Heterogeneous Equilibria

Consider the equilibrium: $H_2O\ (\ell) \rightleftharpoons H_2O(g)$

As we saw in Chapter 10, liquid water has an equilibrium vapor pressure that depends only on the temperature, not on how much liquid is present. The "effective concentration" of the pure liquid water is constant. We can write the equilibrium constant for this reaction:

$$K = [H_2O(g)]$$

This equilibrium constant applies so long as some liquid water is present.

For the reaction: $CaCO_3(s) \rightleftharpoons CaO(s) + CO_2(g)$

the "effective concentrations" of solid $CaCO_3$ and CaO are constant. The equilibrium constant for this reaction is:

$$K = [CO_2(g)]$$

This equilibrium constant applies so long as both CaO and CaO_3 are present.

Rule: Pure solids or pure liquids never appear in the equilibrium expression for a reaction.

The Reaction Quotient

How can we tell whether a reaction mixture is at equilibrium? If the mixture isn't at equilibrium, in which direction does the reaction proceed to reach equilibrium?

We can use the reaction quotient, Q. The reaction quotient has the same form as the equilibrium constant, with actual concentrations in place of the equilibrium concentrations.

Problem 13-4:

At 227°C, the equilibrium constant for the reaction:

$$N_2(g) + 3 H_2(g) \rightleftharpoons 2 NH_3(g) \qquad K = 168$$

If a 2.00 liter container contains 0.100 moles of NH_3, 0.200 moles N_2 and 0.300 moles of H_2, is the mixture at equilibrium? If not, in which direction will reaction take place?

Solution:

Use the reaction quotient, Q.

$$[NH_3] = \frac{0.100 \text{ moles}}{2.00 \text{ L}} = 0.0500 \text{ moles/L}$$

$$[N_2] = \frac{0.200 \text{ moles}}{2.00 \text{ L}} = 0.100 \text{ moles/L}$$

$$[H_2] = \frac{0.300 \text{ moles}}{2.00 \text{ L}} = 0.150 \text{ moles/L}$$

$$Q = \frac{[NH_3]^2}{[N_2][H_2]^3} = \frac{(0.0500)^2}{(0.100)(0.150)^3} = 7.41$$

Q is less than K; that is, the concentrations of N_2 and H_2 are too high, and the concentration of NH_3 is too low. Therefore, to reach the equilibrium position, the reaction must produce NH_3. The reaction proceeds forward.

If Q is equal to K, the system is at equilibrium.

If Q is less than K, the reaction proceeds forward to produce products and to consume reactants.

If Q is greater than K, the reaction must occur in the reverse direction to produce reactants and consume products.

Solving Equilibrium Problems

We need a strategy for solving "equilibrium problems," where we are asked to find equilibrium concentrations, given the equilibrium constant and the initial concentrations. At first, equilibrium problems may seem a bit overwhelming. Don't be fooled by appearances.

The strategy:

• First, write the balanced equation for the equilibrium.

• Then, write the equilibrium expression.

- Use the initial concentrations to calculate Q, the reaction quotient.

- From Q, determine the direction the reaction occurs to get to equilibrium.

- Use x to represent the change in one concentration.

- Use the balanced equation to calculate the other changes in concentration in terms of x.

- Write the equilibrium concentrations in terms of the initial concentrations and the changes.

- Substitute these expressions for the equilibrium concentrations into the equilibrium expression <u>and</u>

- Solve for x.

Problem 13-5:

a. 3.00 moles of HF, 2.00 moles of F_2 and 2.00 moles H_2 are placed in a 2.00 liter flask. At this temperature:

$$H_2(g) + F_2(g) \rightleftharpoons 2\ HF(g)$$

The equilibrium constant is 115. What is the equilibrium concentration of HF in the flask?

Solution:

The balanced equation: $H_2(g) + F_2(g) \rightleftharpoons 2\ HF(g)$

The equilibrium expression:

$$K_{eq} = \frac{[HF]^2_{eq}}{[H_2]_{eq}[F_2]_{eq}} = 115$$

The initial concentrations:

$$[HF] = \frac{3.00\ \text{moles}}{2.00\ \text{liters}} = 1.50\ \text{moles/liter}$$

$$[H_2] = \frac{2.00\ \text{moles}}{2.00\ \text{liters}} = 1.00\ \text{moles/liter}$$

$$[F_2] = \frac{2.00\ \text{moles}}{2.00\ \text{liters}} = 1.00\ \text{moles/liter}$$

The reaction quotient:

$$Q = \frac{[HF]^2}{[H_2][F_2]} = \frac{(1.50)^2}{(1.00)(1.00)} = 2.25$$

Q is less than K: The reaction proceeds forward. H_2 and F_2 react to form HF:

Let x be the moles/L of H_2 that react:

x moles/L of F_2 react and $2x$ moles/L of HF form.

The equilibrium concentration of H_2 is $[H_2]_{eq} = 1.00 - x$.

$[F_2]_{eq} = 1.00 - x$ and $[HF]_{eq} = 1.50 + 2x$

Then:

$$K_{eq} = \frac{[HF]^2}{[H_2][F_2]} = \frac{(1.50 + 2x)^2}{(1.00 - x)(1.00 - x)} = 115$$

Now: solve for x:

$$\frac{(1.50 + 2x)^2}{(1.00 - x)^2} = 115$$

Take the square root of both sides of the equation:

$$\frac{1.50 + 2x}{1.00 - x} = \sqrt{115} = 10.72$$

Rearrange:

$$1.50 + 2x = 10.72(1.00 - x)$$

$$1.50 + 2x = 10.72 - 10.72\ x$$

$$12.72x = 10.72 - 1.50 = 9.22$$

$$x = 0.725$$

Now calculate the equilibrium concentrations:

$$[HF]_{eq} = 1.50 + 2x$$

$$= 1.50 + 2(0.725)$$

$$= 2.95 \text{ moles/liter}$$

$$[H_2] = 1.00 - x$$

$$= 1.00 - 0.725$$

$$= 0.27_5 = 0.28 \text{ moles/liter}$$

b. 3.00 moles of HF, 1.00 mole of F_2 and 2.00 moles of H_2 are placed in a 1.00 liter flask, at the same temperature as part a. What is the equilibrium concentration of HF in this mixture?

Solution:

Use the same balanced equation and equilibrium expression as in part a.

The initial concentrations:

$$[HF] = 3.00 \text{ moles/L}$$

$$[F_2] = 1.00 \text{ moles/L}$$

$$[H_2] = 2.00 \text{ moles/L}$$

$$Q = \frac{[HF]^2}{[H_2][F_2]} = \frac{(3.00)^2}{(2.00)(1.00)} = 4.50$$

Q is less than K: the reaction proceeds forward.

x moles/L of H_2 react, x moles/L of F_2 react, 2x moles/L of HF form. (This seems familiar.)

$$[H_2]_{eq} = 2.00 - x$$

$$[F_2]_{eq} = 1.00 - x$$

$$[HF]_{eq} = 3.00 + 2x$$

$$K_{eq} = \frac{[HF]^2}{[H_2][F_2]} = \frac{(3.00 + 2x)^2}{(2.00 - x)(1.00 - x)} = 115$$

Now, solve for x.

$$(3.00 + 2x)^2 = 115(2.00 - x)(1.00 - x)$$

$$9.00 + x + 4x^2 = 115(2.00 - 3.00x + x^2)$$

$$9.00 + 12.00x + 4x^2 = 230 - 345x + 115x^2$$

Collect terms on one side, to give an equation of the form:

$$ax^2 + bx + c = 0.$$

$$111x^2 - 357x + 221 = 0$$

Use the quadratic formula to find values of x which satisfy the equation:

$$x = \frac{-b \pm \sqrt{b^2 - 4ac}}{2a}$$

$$= \frac{357 \pm \sqrt{(357)^2 - 4(111)(221)}}{2(111)}$$

$$= \frac{357 \pm \sqrt{29325}}{222}$$

$$= \frac{357 \pm 171}{222}$$

There are two possible values of x

$$x = \frac{357 + 171}{222} \qquad\qquad x = \frac{357 - 171}{222}$$

$$x = 2.38 \qquad\qquad x = .838$$

Which answer makes sense? If we choose $x = 2.38$, the equilibrium concentration of F_2 would be negative. Negative concentrations are impossible. Then $x = .838$.

$$[HF]_{eq} = 3.00 + 2x = 3.00 + 1.68 = 4.68 \text{ mol/L}$$

Approximations

At high temperatures N_2 and O_2 can react to form NO

$$N_2(g) + O_2(g) \rightleftharpoons 2\,NO(g)$$

At a certain temperature, $K_c = 3.6 \times 10^{-4}$.

What is the equilibrium concentration of NO when 0.100 moles of N_2 and 0.200 moles of O_2 are placed in a 2.00 liter container?

	N_2	O_2	NO
Initial concentrations	0.0500 moles/L	0.100 moles/L	0 moles/L
Final concentrations	0.0500 − x	0.100 − x	2x

$$K_c = \frac{[NO]^2}{[N_2][O_2]} = \frac{(2x)^2}{(0.0500 - x)(0.100 - x)} = 3.6 \times 10^{-4}$$

We could solve the resulting quadratic equation exactly. However, if we are clever we can make some assumptions, and simplify the mathematics. (We do have to remember to check to see if we've made valid assumptions.)

In this case, K_c is much less than 1. Let's assume that very little NO forms. Then the equilibrium concentrations of N_2 and O_2 are very nearly equal to the initial concentrations; that is x is small compared to 0.0500 and 0.100, and 0.0500 − x is very nearly 0.0500, and 0.100 − x is very nearly 0.100.

Then

$$K_c = \frac{(2x)^2}{(0.0500 - x)(0.100 - x)} = 3.6 \times 10^{-4}$$

Becomes

$$\frac{(2x)^2}{(0.0500)(0.100)} = 3.6 \times 10^{-4}$$

Solving for x:

$$4x^2 = 1.8 \times 10^{-6}$$

$$x^2 = 4.5 \times 10^{-7}$$

$$x = 6.7 \times 10^{-4}$$

x is small compared to both 0.100 and 0.0500. How small is small enough? If x is less than 1/20th (0.05) of a number, it can be ignored when added to or subtracted from that number.

$$[NO] = 2x = 1.3 \times 10^{-3} \ M$$

Le Chatelier's Principle

We can make qualitative predictions about how a system at equilibrium will respond to changes in concentrations, changes in pressure and changes in temperature.

When we impose a change on a chemical system at equilibrium, the system will respond to counteract the change (if it can).

If we add a reactant (or product) to a chemical system at equilibrium, the reaction will occur to consume part of the added material, until the system reaches a new equilibrium position. If we add NH_3 to an equilibrium mixture of N_2, H_2 and NH_3, some of the NH_3 will react to form N_2 and H_2 until equilibrium is reestablished. The equilibrium constant for the reaction doesn't change; the system reaches a new equilibrium position. If we increase the pressure on a chemical system at equilibrium, the reaction will occur to reduce the total pressure of the system, if it can. The equilibrium N_2 + 3 H_2 ⇌ 2 NH_3 will "shift" to form NH_3 as pressure is increased. The reaction produces NH_3, reducing the total number of moles of gas in the mixture. Increasing the pressure doesn't change the equilibrium constant for the reaction. Increasing the pressure does change the concentrations of gases. The reaction proceeds until the system reaches a new equilibrium mixture.

If we raise the temperature of a chemical system at equilibrium, the reaction will proceed to consume some of the added energy. The reaction N_2 + 3 H_2 ⇌ 2 NH_3 is exothermic, that is, heat is produced as NH_3 forms. If heat is added to raise the temperature of an equilibrium mixture of N_2, H_2 and NH_3, NH_3 reacts to form more N_2 and H_2, consuming some of the added energy. When the temperature is changed, the equilibrium constant does change.

Problem 13-6:

The reaction of H_2 with I_2 to form HI is exothermic.

$$H_2(g) + I_2(g) \rightleftharpoons 2\ HI(g) \qquad \Delta H_{rxn} = -10\ kJ$$

a. If the temperature of an equilibrium mixture of H_2, I_2 and HI is raised, does the concentration of HI increase or decrease?

b. If the pressure is increased on this mixture, does more HI form?

Solution:

a. We could write the reaction as:

$$H_2(g) + Br_2(g) \rightleftharpoons 2\ HBr(g) + heat\ (energy)$$

If heat is added, the reaction will shift to consume part of the added energy. HBr will react to form more H_2 and more Br_2.

b. As pressure on a system at equilibrium is increased, a reaction will shift to form fewer moles of gas. In this reaction, two moles of gaseous reactants form two moles of gaseous product. This reaction has no way to counteract the increased pressure. The number of moles of HBr remains the same when the pressure is increased.

CHAPTER THIRTEEN: SELF-TEST

13-1. Write the equilibrium expression for each of the following reactions:

a. $N_2O_4(g) \rightleftharpoons 2\ NO_2(g)$

b. $N_2(g) + 3\ H_2(g) \rightleftharpoons 2\ NH_3(g)$

13-2. An equilibrium mixture of N_2, H_2 and NH_3 is measured at a particular temperature.

Mixture	$[NH_3]$	$[N_2]$	$[H_2]$
i)	0.050 moles/liter	0.20 moles/liter	0.60 moles/liter

a. Calculate K_{eq} for the reaction:

$$N_2(g) + 3\ H_2(g) \rightleftharpoons 2\ NH_3(g)$$

b. Calculate K_{eq} for the reaction:

$$2\ NH_3(g) \rightleftharpoons N_2(g) + 3\ H_2(g)$$

c. What concentration of NH_3 is in equilibrium with 0.30 moles/liter N_2 and 0.50 moles/liter H_2 at this temperature?

13-3. At a certain temperature, the equilibrium constant for the reaction

$$2\ SO_2(g) + O_2(g) \rightleftharpoons 2\ SO_3(g)$$

is: $K_c = 2.80 \times 10^2$

If a mixture of SO_2, O_2, and SO_3 has the following concentrations:

$[SO_2] = 0.10$ moles/L

$[O_2] = 0.20$ moles/L

$[SO_3] = 0.30$ moles/L

In which direction will the reaction go to reach equilibrium?

13-4. 0.250 moles of NH_3 is placed in a 2.00 liter container at a particular temperature. NH_3 reacts to produce N_2 and H_2: $2\ NH_3(g) \rightleftharpoons N_2(g) + 3\ H_2(g)$. At equilibrium 0.150 moles of NH_3 remains.

a. What is the equilibrium concentration of N_2?

b. What is the equilibrium concentration of H_2?

c. Calculate K_{eq} for the reaction at this temperature.

13-5.   At a particular temperature, the equilibrium constant for the reaction:

$$H_2(g) + F_2(g) \rightleftharpoons 2 \; HF(g) \quad K_c = 6.40$$

If 3.00 moles of HF is placed in a 2.00 liter container,

a.   What is the equilibrium concentration of H_2 in the flask?

b.   What is the equilibrium concentration of HF in the flask?

Hint: Complete the concentration table:

	$[H_2]$	$[F_2]$	$[HF]$
Initially	0	0	—
Reacted	—	—	—
Finally	—	—	—

13-6.   At high temperatures N_2 and O_2 can react to form NO:

$$N_2(g) + O_2(g) \rightleftharpoons 2 \; NO(g)$$

At a certain temperature $K_c = 3.6 \times 10^{-4}$.

0.20 moles of N_2 and 0.30 moles of O_2 are placed in a 1.00 L container.

a.   What is the equilibrium concentration of NO?

b.   What is the equilibrium concentration of N_2?

Hint: Set up the concentration table. Use the equilibrium expression to solve for x.

The next three problems refer to an equilibrium mixture of NO_2 and N_2O_4 in a container:

$$N_2O_4(g) \rightleftharpoons 2 \; NO_2(g) \quad \Delta H = +57 \; kJ$$

13-7. $NO_2(g)$ is added to the equilibrium mixture in the container, and a new equilibrium position is reached.

a.   How does the concentration of N_2O_4 change?

b.   How does the equilibrium constant change?

13-8.   The volume of the container is decreased, and a new equilibrium position is reached.

a.   In which direction does the reaction take place?

b.   How does the equilibrium constant change?

13-9. The temperature of the mixture is increased, and a new equilibrium
position is reached.

    a.   In which direction does the reaction take place?

    b.   How does the equilibrium constant change?

CHAPTER THIRTEEN: SELF-TEST ANSWERS

13-1. a. $K_{eq} = \dfrac{[NO_2]^2}{[N_2O_4]}$

b. $K_{eq} = \dfrac{[NH_3]^2}{[N_2][H_2]^3}$

13-2. a. $K_{eq} = 5.8 \times 10^{-2}$

b. $K_{eq} = 17$

c. 0.047 moles/liter

13-3. $Q = \dfrac{(0.30)^2}{(0.10)^2(0.20)} = 45$

Q is less than K_{eq}: the reaction will proceed forward.

13-4. 0.100 moles NH_3 react to form 0.0500 moles N_2 and 0.150 moles H_2; $[NH_3] = 7.50 \times 10^{-2}$ moles/L

a. $[N_2] = 2.50 \times 10^{-2}$ moles/L

b. $[H_2] = 7.50 \times 10^{-2}$ moles/L

c. $K = \dfrac{(7.50 \times 10^{-2})^2}{(2.5 \times 10^{-2})(7.5 \times 10^{-2})^3} = 5.3 \times 10^2$

13-5.

	$[H_2]$	$[F_2]$	[HF]
Initially	0	0	1.50 moles/liter
Reacted			2x
Finally	x	x	1.50 - 2x

$K_{eq} = \dfrac{(1.50 - 2x)^2}{x^2} = 6.40$

$\dfrac{(1.50 - 2x)}{x} = \sqrt{6.40} = 2.53$

a. $[H_2] = x = 0.331$ moles/L

b. [HF] = 1.50 moles/L - 2x = 0.84 moles/L

13-6.

	[NO]	$[N_2]$	$[O_2]$
Initially	0	0.20 moles/L	0.30 moles/L
Reacted		x	x
Finally	$2x$	$0.20 - x$	$0.30 - x$

$$K = \frac{(2x)^2}{(0.20 - x)(0.30 - x)} = 3.6 \times 10^{-4}$$

$$\frac{(2x)^2}{(0.20)(0.30)} \approx 3.6 \times 10^{-4}$$

$$x = 2.3 \times 10^{-3} \text{ moles/L}$$

a. $[NO] = 4.6 \times 10^{-3}$ moles/L

b. $[N_2] = (0.20 - x)$ moles/L $= 0.20$ moles/L

13-7. a. The concentration of N_2O_4 increases.

b. The equilibrium <u>constant</u> doesn't change.

13-8. a. More N_2O_4 forms.

b. The equilibrium <u>constant</u> doesn't change.

13-9. a. The reaction is endothermic; more NO_2 forms.

b. The equilibrium <u>constant</u> increases.

In this chapter we meet three different definitions of acids and bases.

	Arrhenius	Brönsted–Lowry	Lewis
Acid	Produces H^+ in solution	Proton donor	Electron pair acceptor
Base	Produces OH^- in solution	Proton acceptor	Electron pair donor

Look at the reaction of acetic acid in water:

$$CH_3 - \overset{\displaystyle O}{\overset{\displaystyle \|}{C}} - O - H + H_2O \rightleftharpoons H_3O^+ + CH_3 - \overset{\displaystyle O}{\overset{\displaystyle \|}{C}} - O^-$$

H_2O is a Brönsted base, accepting a proton from acetic acid. If we look at the reverse reaction, H_3O^+ is the Brönsted acid, donating a proton to CH_3COO^- (acetate), a Brönsted base.

The Brönsted model looks at conjugate acid–base pairs: H_3O^+ is the conjugate acid of H_2O; H_2O is the conjugate base of H_3O^+. An acid–base reaction can be looked at as a competition between two Brönsted bases for a proton. In this reaction, water molecules and acetate ions are competing for the proton.

$$CH_3 - \overset{\displaystyle O}{\overset{\displaystyle \|}{C}} - O^- - - - H^+ - - - \overset{\displaystyle H}{\overset{\displaystyle |}{O}} - H$$

The stronger base "wins" the competition. K_{eq} for this reaction is 1.8×10^{-5}. The equilibrium favors the reactants. Then acetate must win the competition for the proton: acetate is a stronger base than water.

H_3O^+ loses its proton more easily than CH_3COH does: H_3O^+ is the stronger acid. The equilibrium favors the weaker base (H_2O) and the weaker acid ($CH_3 - \underset{\displaystyle \underset{\displaystyle O}{\|}}{C} - OH$).

Problem 14-1:

$$HNO_2 + CH_3-\overset{\displaystyle O}{\overset{\displaystyle \|}{C}}O^- \rightleftharpoons NO_2^- + CH_3\overset{\displaystyle O}{\overset{\displaystyle \|}{C}}OH \qquad K_{eq} = 22.2$$

a. What are the two acids in this reaction?

b. What are the two bases in this reaction?

c. Which of the two bases is stronger?

d. Which of the two acids is weaker?

a. Acids are proton donors: HNO_2 and CH_3COH (with $\overset{\overset{\displaystyle O}{\displaystyle \|}}{}$) are the two acids.

b. Bases are proton acceptors: CH_3CO^- (with $\overset{\overset{\displaystyle O}{\displaystyle \|}}{}$) and NO_2^- are the two bases.

c & d. K_{eq} is much greater than 1. The equilibrium favors CH_3CO_2H and NO_2^-. CH_3CO_2H must be the weaker acid. NO_2^- must be the weaker base. HNO_2 is the stronger acid. $CH_3CO_2^-$ is the stronger base.

Acid "Dissociation" Constants

When HOCl is dissolved in water, the following reaction occurs:

$$HOCl(aq) + H_2O(\ell) \rightleftharpoons H_3O^+(aq) + OCl^-(aq) \qquad K_{eq} = \frac{[H_3O^+][OCl^-]}{[HOCl]} = 3.5 \times 10^{-8}$$

The reaction favors the reactants. HOCl is a weaker acid than H_3O^+. Water doesn't appear in this equilibrium expression because it is the solvent. Chemists often abbreviate the hydrated hydrogen ion as H^+. Then the abbreviated equilibrium expression is:

$$K_{eq} = \frac{[H^+][OCl^-]}{[HOCl]}$$

The "abbreviated" reaction is: $HOCl \rightleftharpoons H^+ + OCl^-$. This reaction focusses on the "dissociation" of HOCl. The role of water in the reaction is "hidden."

Acid Strengths

We can determine the relative strengths of acids by comparing their "dissociation" constants.

$$HOCl(aq) + H_2O(\ell) \rightleftharpoons OCl^-(aq) + H_3O^+(aq) \qquad K = \frac{[H_3O^+][OCl^-]}{[HOCl]} = 3.5 \times 10^{-8}$$

$$HNO_2(aq) + H_2O(\ell) \rightleftharpoons NO_2^-(aq) + H_3O^+(aq) \qquad K = \frac{[H_3O^+][NO_2^-]}{[HNO_2]} = 4.0 \times 10^{-4}$$

The second reaction favors products more than the first reaction does. The equilibrium constants show that it is easier for H_2O to remove a proton from HNO_2 than from HOCl. HNO_2 is the stronger acid.

Problem 14-2:

Use the data in Table 14-2 in the text to arrange the following acids in order of increasing acid strength.

$$HCN, \ HClO_2, \ NH_4^+, \ HF, \ HNO_2$$

Solution:

Find K_a for each acid:

	K_a
HCN	6.2×10^{-10}
$HClO_2$	1.2×10^{-2}
NH_4^+	5.6×10^{-10}
HF	7.2×10^{-4}
HNO_2	4.0×10^{-4}

The strongest acid has the largest K_a; the weakest acid has the smallest K_a.

In order of increasing acid strength we have:

$$NH_4^+ < HCN < HNO_2 < HF < HClO_2$$

Problem 14-3:

Which is the stronger base: NH_3 or NO_2^-?

Solution:

HNO_2 is a stronger acid than NH_4^+. Therefore, NO_2^-, the conjugate base of HNO_2, is a weaker base than NH_3, the conjugate base of NH_4^+.

Water

Water can act as a proton donor (an acid) and as a proton acceptor (a base).

$$2\ H_2O(\ell) \rightleftharpoons H_3O^+(aq) + OH^-(aq) \qquad K_w = [H_3O^+][OH^-]$$

$$\text{or } K_w = [H^+][OH^-]$$

The equilibrium constant for this reaction is 1.0×10^{-14} at 25°C. In pure water, the concentration of hydrogen ions must equal the concentration of hydroxide ions. In pure water, $[H^+] = [OH^-]$. Then, $K_w = [H^+][OH^-] = [H^+]^2 = 1.0 \times 10^{-14}$. In pure water, at 25°C, $[H^+] = \sqrt{1.0 \times 10^{-14}} = 1.0 \times 10^{-7}\ M$.

K_w applies to any aqueous solution at 25°C. If we know the hydrogen ion concentration, we can calculate the hydroxide ion concentration and vice versa. If the hydrogen ion concentration is larger than the hydroxide ion concentration, the solution is acidic. If the hydroxide ion concentration is larger than the hydrogen ion concentration, the solution is basic. If the hydroxide and hydrogen ion concentrations are equal, the solution is neutral.

Logarithms

We can use a "logarithmic" scale to describe hydrogen ion concentrations. Remember, a logarithm is an exponent. We will use "common" logarithms, logarithms "to the base 10." Let's look at some examples:

$$\log 10^3 = 3$$

$$\log 10^{-2} = -2$$

$$\log 10^{-2.3} = -2.3$$

$$\log 6.5 \times 10^{-2} = -1.19$$

(Use your log key on your calculator.)

Numbers smaller than 1 have negative logarithms. Because the hydrogen ion concentration in most aqueous solutions is small, we will use the <u>pH scale</u>. $pH = -\log [H^+]$.

Problem 14-4:

The hydrogen ion concentration in a solution is 5.0×10^{-8}. What is the pH?

Solution:

$$pH = -\log [H^+] = -\log (5.0 \times 10^{-8})$$

Enter 5.0×10^{-8} in your calculator; then use the log key to take the logarithm; finally use your "change sign" key (CHS or +/-).

$$pH = -\log (5.0 \times 10^{-8})$$

$$= -(-7.30) = 7.30$$

For all aqueous solutions at 25°C, $K_w = [H^+][OH^-] = 1.0 \times 10^{-14}$.

$$\log K_w = \log [H^+] + \log [OH^-] = -14$$

$$-\log K_w = -\log [H^+] -\log [OH^-] = + 14$$

If we define pK_w as $-\log K_w$, and pOH as $-\log [OH^-]$, then

$$pK_w = pH + pOH = 14.$$

In a neutral solution: $[H^+] = [OH^-]$

Then: $pH = pOH = 7$

An acidic solution has a pH less than 7; a basic solution has a pH greater than 7.

Problem 14-5:

The pH of a solution is 5.70. What is the hydrogen ion concentration of this solution?

Solution:

$$pH = 5.70 = -\log [H^+]$$

$$\log [H^+] = -5.70$$

$$\text{or} \quad [H^+] = 10^{-5.70}$$

First: Enter 5.70

Then: Change the sign: -5.70

Use your antilog (inv log) or 10^x key: $1.995.... \times 10^{-6}$

If the pH has two decimal places, the answer has two significant figures. For this solution, then, $[H^+] = 2.0 \times 10^{-6}$ M.

Acid–Base Equilibrium Problems

We'll use a systematic approach to these problems, "thinking each one out."

1. First, list all the species in the solution.

2. Then, write all the equilibria. Find the dominant (most important) equilibrium, the equilibrium with the largest equilibrium constant.

3. Next, determine the major species in solution.

4. Then, you are ready to begin using the equilibrium expressions, constructing a concentration table.

For example, calculate the hydrogen ion concentration and pH of a 0.500 M acetic acid solution. (CH_3CO_2H)

$$
\text{All the species:} \quad CH_3\overset{\displaystyle O}{\overset{\displaystyle \|}{C}} - O - H, \quad CH_3\overset{\displaystyle O}{\overset{\displaystyle \|}{C}} - O^-, \quad H^+, \quad OH^-, \quad H_2O
$$

The equilibria: $H_2O \rightleftharpoons H^+ + OH^- \quad K_w = 1.0 \times 10^{-14}$

$$
CH_3COH \rightleftharpoons CH_3\overset{\displaystyle O}{\overset{\displaystyle \|}{C}} - O + H^+ \quad K_a = 1.8 \times 10^{-5}
$$

The dissociation of acetic acid is the major equilibrium. There are two sources of hydrogen ions in solution, the dissociation of acetic acid, and the ionization of water. Because K_a is so much larger than K_w, we'll assume that acetic acid is the principal source of hydrogen ions in solution. (Remember to check our assumption later.)

The major species are H_2O, $CH_3\overset{\overset{\displaystyle O}{\displaystyle\|}}{C}OH$, $CH_3\overset{\overset{\displaystyle O}{\displaystyle\|}}{C}O^-$ and H^+.

If all (or nearly all) of the hydrogen ions come from the dissociation of acetic acid, we can set up a table.

	$[H^+]$	$[CH_3\overset{\overset{\displaystyle O}{\displaystyle\|}}{C}O^-]$	$[CH_3\overset{\overset{\displaystyle O}{\displaystyle\|}}{C}OH]$
Initial concentration	~0	0	0.500
Final concentration	x	x	$0.500 - x$

Use K_a:

$$K_a = \frac{[H^+][CH_3COO^-]}{[CH_3COOH]} = 1.8 \times 10^{-5}$$

$$\frac{x \cdot x}{0.500 - x} = 1.8 \times 10^{-5}$$

Now we have a choice: we can solve the resulting quadratic equation or we can use our intuition, and assume that only a small fraction of the acetic acid dissociates. Then $0.500 - x$ is approximately 0.500.

$$\frac{x^2}{0.500} = 1.8 \times 10^{-5}$$

$$x^2 = 9.0 \times 10^{-6}$$

$$[H^+] = x = 3.0 \times 10^{-3} \; M$$

Let's check our assumptions:

- Is the amount of acetic acid that dissociates negligible? Using our "5% rule," we find that $x/0.500 = 0.006$. Because this is less than 0.05 (x is less than 5% of 0.500), our assumption is valid.

- Is the acetic acid the principal source of hydrogen ions? If we calculate the concentration of hydroxide ions, we will have the concentration of hydrogen ions produced by the ionization of water.

$$H_2O(\ell) \rightleftharpoons H^+(aq) + OH^-(aq)$$

$$K_w = [H^+][OH^-] = 1.0 \times 10^{-14}$$

If the total $[H^+] = 3.0 \times 10^{-3} \; M$

$$[OH^-] = \frac{1.0 \times 10^{-14}}{3.0 \times 10^{-3}} = 3.3 \times 10^{-12} \; M$$

For each hydroxide ion present in this solution, there was a hydrogen ion produced.

We could think of the total hydrogen ion concentration as the sum of the hydrogen ion concentration produced by dissociation of the acetic acid and the hydrogen ion concentration produced by the ionization of water.

$$[H^+]_{total} = [H^+]_{acetic\ acid} + [H^+]_{H_2O}$$

In this solution:

$$3.0 \times 10^{-3}\ M = [H^+]\ acetic\ acid$$

$$[H^+]_{total} = 3.3 \times 10^{-12}\ M$$

Acetic acid is the principal source of hydrogen ion in this solution. We can ignore water as a significant source of H^+.

Problem 14-6:

What is the pH of an aqueous $1.00 \times 10^{-9}\ M$ HCl solution?

Solution:

All the species: H^+, OH^-, Cl^-, H_2O

The equilibria: $HCl(aq) \rightarrow H^+(aq) + Cl^-(aq)$ K is very large

$$H_2O(\ell) \rightleftharpoons H^+(aq) + OH^-(aq)\qquad K_w = 1.0 \times 10^{-14}$$

HCl is a strong acid and dissociates completely.

Which species are the major species? In pure water, $[H^+] = 1.0 \times 10^{-7}\ M$. If we add 1.00×10^{-9} moles of HCl to 1.00 liter of pure water, water is still the principal source of hydrogen ions. The pH of this solution can't be above 7.00.

(If we hadn't been careful, we might have jumped to the conclusion that the pH of this solution was 9. The systematic approach keeps us from going astray.)

In this solution:

$$[H^+] = [OH^-] + [Cl^-]$$

The chloride ion concentration is small enough to be ignored.

$$[H^+] = [OH^-] = x$$

$$[H^+][OH^-] = x^2 = 1.0 \times 10^{-14}$$

$$[H^+] = 1.0 \times 10^{-7}\ M$$

216

Dilution

Problem 14-7:

    a. What is the pH of a 0.100 M CH_3COOH solution ($K_a = 1.8 \times 10^{-5}$)?

    b. What is the pH of a solution made by adding 90.0 mL of water to 10.0 mL of this acetic acid solution?

Solution:

    a. All species: CH_3COOH, CH_3COO^-, H^+, OH^-, H_2O

The equilibria:

$$CH_3COOH(aq) \rightleftharpoons CH_3COO^-(aq) + H^+(aq) \quad K_a = 1.8 \times 10^{-5}$$

$$H_2O(\ell) \rightleftharpoons H^+(aq) + OH^-(aq) \quad K_w = 1.0 \times 10^{-14}$$

The major equilibrium: $CH_3COOH(aq) \rightleftharpoons CH_3COO^-(aq) + H^+(aq)$, because K_a is much larger than K_w, <u>and</u> the acetic acid solution is not very dilute.

The major species: CH_3COOH, CH_3COO^-, H^+, H_2O

Now construct a table:

	CH_3COOH	CH_3COO^-	H^+
Initial concentration	0.100	0	~0
Final concentration	$0.100 - x$	x	x

Use the equilibrium expression:

$$K_a = \frac{[H^+][CH_3COO^-]}{[CH_3COOH]} = \frac{x^2}{0.100 - x} = 1.8 \times 10^{-5}$$

Assume x is small compared to 0.100:

$$\frac{x^2}{0.100} = 1.8 \times 10^{-5}$$

$$x^2 = 1.8 \times 10^{-6}$$

$$[H^+] = x = \sqrt{1.8 \times 10^{-6}} = 1.3 \times 10^{-3} \ M$$

(Check: Is $1.3 \times 10^{-3} < 0.05(0.100)$? Yes.)

$$pH = -\log[H^+] = -\log(1.3 \times 10^{-3})$$

$$= -(-2.89)$$

$$= 2.89$$

b. Everything is the same except the table: the initial concentration of CH_3COOH in the dilute solution is 0.0100 M. (Remember dilution problems?)

	CH_3COOH	CH_3COO^-	H^+
Initial concentration	0.0100	0	~0
Final concentration	$0.0100 - x$	x	x

$$K_a = \frac{x^2}{0.0100 - x} = 1.8 \times 10^{-5}$$

$$x^2 = 1.8 \times 10^{-7}$$

$$[H^+] = x = \sqrt{1.8 \times 10^{-7}}$$

$$[H^+] = 4.2 \times 10^{-4}$$

(Check: Is $4.2 \times 10^{-4} < 0.05(0.0100)$? Just barely.)

$$pH = 3.38$$

Percent Dissociation

Problem 14-8:

a. What percent of the acetic acid is dissociated in 0.100 M CH_3COOH?

b. In 0.0100 M CH_3COOH?

$$\text{percent dissociation} = \frac{\text{amount dissociated}}{\text{total}} \times 100 = \frac{[CH_3COO^-]}{\text{total}} \times 100$$

a. From 14-7: In these solutions the concentration of acetate ion equals the concentration of hydrogen ion. In 0.100 M CH_3COOH:

$$[CH_3COO^-] = [H^+] = 1.3 \times 10^{-3} \ M$$

$$\text{percent dissociated} = \frac{1.3 \times 10^{-3}}{0.100} = 1.3\%$$

b. In 0.0100 M CH_3COOH:

$$[CH_3COO^-] = [H^+] = 4.2 \times 10^{-4} \ M$$

$$\text{percent dissociation} = \frac{4.2 \times 10^{-4}}{0.0100} \times 100 = 4.2\%$$

As we dilute the acetic acid, a greater fraction of the acid dissociates.

Bases

We can use the same approach to equilibria for bases as we did for acids.

Problem 14-9:

What is the pH of a 4.5×10^{-2} M $Ca(OH)_2$?

Solution:

$Ca(OH)_2$ is a strong base, dissociating completely:

All the species: Ca^{2+}, OH^-, H_2O, H^+.

The equilibrium: $H_2O(\ell) \rightleftharpoons H^+ + OH^-(aq)$ $K_w = 1.0 \times 10^{-14}$

The major species: the hydroxide ion from $Ca(OH)_2$ "suppresses" the ionization of water. In pure water the concentration of hydrogen ion is 1.0×10^{-7} M. In this solution the hydrogen ion concentration is even lower.

The major species, then, are: Ca^{2+}, OH^- and H_2O.

$$[Ca^{2+}] = 4.5 \times 10^{-2} \ M$$

$$[OH^-] = 2 \ [Ca^{2+}] = 9.0 \times 10^{-2} \ M$$

$$[H^+] = \frac{1.0 \times 10^{-14}}{[OH^-]} = \frac{1.0 \times 10^{-14}}{9.0 \times 10^{-2}} = 1.1 \times 10^{-13} \ M$$

$$pH = -\log \ [H^+] = -\log \ (1.1 \times 10^{-13} = -[-12.96] = 12.96$$

We could have calculated the pH another way:

$$[OH^-] = 9.0 \times 10^{-2} \ M$$

$$pOH = -\log \ [OH^-] = 1.04$$

$$pH = 14 - pOH = 14 - 1.04 = 12.96$$

Problem 14-10:

What is the pH of a 0.0500 M methylamine (CH_3NH_2) solution?

All the species: CH_3NH_2, $CH_3NH_3^+$, OH^-, H^+, H_2O.

The equilibria:

$$CH_3NH_2(aq) + H_2O(\ell) \rightleftharpoons CH_3NH_3^+(aq) + OH^-(aq) \quad K_b = 4.38 \times 10^{-4}$$

$$H_2O \rightleftharpoons H^+ + OH^- \quad K_w = 1.0 \times 10^{-14}$$

The principal equilibrium is the dissociation of methyl amine: K_b is much larger than K_w.

Then, the principal species are: CH_3NH_2, $CH_3NH_3^+$, OH^-, and H_2O.

$$K_b = \frac{[CH_3NH_3^+][OH^-]}{[CH_3NH_2]} = 4.38 \times 10^{-4}$$

Construct a concentration table:

	$[CH_3NH_2]$	$[CH_3NH_3^+]$	$[OH^-]$
Initial	0.0500	0	~0
Final	0.0500 – x	x	x

$$K_b = \frac{x^2}{0.0500 - x} = 4.38 \times 10^{-4}$$

Again, we could rearrange this expression and solve the quadratic equation, or, we can use a simplifying approximation. Let's assume that x is small compared to 0.050; that is, let's assume that only a small amount of methylamine reacts with water.

$$K_b \simeq \frac{x^2}{.0500} = 4.38 \times 10^{-4}$$

$$x^2 = 2.19 \times 10^{-5}$$

$$x = 4.68 \times 10^{-3}$$

Let's test our assumption: Is x less than 5% of 0.050?

$$\frac{x}{0.0500} = \frac{4.68 \times 10^{-3}}{0.0500} = .094$$

Rats! x is not small enough to ignore. We can go back to our quadratic equation, or, we can use our value x to make a new estimate of the concentration of CH_3NH_2.

$$[CH_3NH_2] = 0.0500 - x \simeq 0.0500 - 4.68 \times 10^{-3} = 4.53 \times 10^{-2}$$

Solve for a new estimate of x, x':

$$K_b \simeq \frac{(x'^2)}{4.53 \times 10^{-2}} = 4.38 \times 10^{-4}$$

$$(x')^2 = (4.38 \times 10^{-4})(4.53 \times 10^{-2}) = 1.98 \times 10^{-5}$$

$$x' = 4.45 \times 10^{-3} \ M$$

Our new estimate of $[CH_3NH_2]$ is: $0.0500 = 4.45 \times 10^{-3} = 4.55 \times 10^{-2} \ M$.

If we use this concentration of CH_3NH_2 to calculate a new estimate of x, x",

$$\frac{(x'')^2}{4.55 \times 10^{-2}} = 4.38 \times 10^{-4}$$

$$(x'') = (4.38 \times 10^{-4})(4.55 \times 10^{-2})$$

$$x'' = 4.46 \times 10^{-3} \ M$$

This is the method of "successive approximations." x'' is very nearly the same as x': We'll take x'' as our answer. We can check by substituting x'' into the expression for K_b.

$$\frac{(4.46 \times 10^{-3})^2}{(.0500 - 4.46 \times 10^{-3})} = 4.37 \times 10^{-4} = K_b$$

This calculated value of K_b is very close to the given value: therefore x'' is the solution to our equation.

Polyprotic Acids

Look at the step-wise dissociation constants for polyprotic acids in Table 14.4. Successive dissociation constants usually differ by at least a factor of 1000. This often allows us to use one principal equilibrium in pH calculations.

Problem 14-11:

Calculate the hydrogen ion concentration in a 0.0500 M ascorbic acid ($H_2C_6H_6O_6$) solution. Ascorbic acid is a diprotic acid.

Solution:

All the species: $H_2C_6H_6O_6$, $HC_6H_6O_6^-$, $C_6H_6O_6^{2-}$, H^+, OH^-, H_2O

The equilibria:

$$H_2C_6H_6O_2(aq) \rightleftharpoons HC_6H_6O_6^-(aq) + H^+(aq) \qquad K_{a_1} = 7.9 \times 10^{-5}$$

$$HC_6H_6O_6^-(aq) \rightleftharpoons C_6H_6O_6^{2-}(aq) + H^+(aq) \qquad K_{a_2} = 1.6 \times 10^{-12}$$

$$H_2O(\ell) \rightleftharpoons H^+(aq) + OH^-(aq) \qquad K_w = 1.0 \times 10^{-14}$$

The principal equilibrium is the first dissociation of ascorbic acid. We'll assume that the concentration of hydrogen ions produced by the second dissociation of ascorbic acid and by the ionization of water is negligible. We'll check these assumptions later.

Set up a concentration table:

	$[H_2C_6H_6O_6]$	$[HC_6H_6O_6^-]$	$[H^+]$
Initial	0.0500	0	~0
Final	0.0500 − x	x	x

Use K_{a_1}:

$$\frac{[H^+][HC_6H_6O_6^-]}{[H_2C_6H_6O_6]} = \frac{x^2}{0.0500 - x} = 7.9 \times 10^{-5}$$

Let's assume that x is small compared to 0.0500:

$$\frac{x^2}{.0500} = 7.9 \times 10^{-5}$$

$$[H^+] = x = 2.0 \times 10^{-3}$$

$$pH = -\log (2.0 \times 10^{-3} = 2.70$$

Let's check the assumptions:

First, is $x/0.050$ less than 0.05? (Is x less than 5% of 0.050 M?)

$$\frac{2.0 \times 10^{-3}}{0.050} = .04$$

The assumption that x is small compared to 0.050 is valid.

Next, is the hydrogen ion concentration due to the dissociation of water negligible? (There is a hydrogen ion produced for every hydroxide ion.) What is the equilibrium concentration of hydroxide ions in this solution? We'll use our calculated $[H^+]$ and K_w:

$$[OH^-] = \frac{K_w}{[H^+]} = \frac{1.0 \times 10^{-14}}{2.0 \times 10^{-3}} = 5.0 \times 10^{-12} \; M$$

The dissociation of water produces 5.0×10^{-12} moles/liter of hydrogen ions in this solution. This is negligible compared to 2.0×10^{-3} moles/liter.

Finally, is the hydrogen ion concentration due to the second dissociation of ascorbic acid negligible? There is a hydrogen ion produced for every $C_6H_6O_6^{2-}$ ion. What is the equilibrium concentration of $C_6H_6O_6^{2-}$ ions in this solution? We'll use our calculated $[H^+]$, $[HC_6H_6O_6^-]$, and K_{a_2}.

$$\frac{[H^+][C_6H_6O_6^{2-}]}{[HC_6H_6O_6^-]} = K_{a_2} = 1.6 \times 10^{-12}$$

In this solution:

$$[H^+] = 2.0 \times 10^{-3} \; M \text{ and}$$

$$[HC_6H_6O_6^-] = 2.0 \times 10^{-3} \; M$$

$$[C_6H_6O_6^{2-}] = \frac{[HC_6H_6O_6^-]}{[H^+]} K_{a_2} = \frac{2.0 \times 10^{-3}}{2.0 \times 10^{-3}} (1.6 \times 10^{-12})$$

$$= 1.6 \times 10^{-12} \; M$$

The second dissociation of ascorbic acid produces 1.6×10^{-12} moles/liter of hydrogen ions in this solution. This is negligible compared to 2.0×10^{-3} moles/liter.

Problem 14-12:

Calculate the pH of a 0.50 M H_2SO_4 solution.

Solution:

H_2SO_4 dissociates completely to give HSO_4^- and H^+. HSO_4^- is itself a weak acid.

The principal species are: H^+, HSO_4^-, and H_2O.

We can set up a concentration table:

	$[H_2SO_4]$	$[HSO_4^-]$	$[SO_4^{2-}]$	$[H^+]$
Initial	0.50	0	0	~0
After the first dissociation	0	0.50	0	0.50
At equilibrium	0	0.50 - x	x	0.50 + x

Then we use K_{a_2}:

$$K_{a_2} = \frac{[H^+][SO_4^{2-}]}{[HSO_4^-]} = \frac{(0.50 + x)(x)}{(0.50 - x)} = 1.2 \times 10^{-2}$$

Let's assume that x is small compared to 0.50:

Then:

$$K_{a_2} = \left(\frac{0.50}{0.50}\right)x = 1.2 \times 10^{-2}$$

$$x = 1.2 \times 10^{-2}\ M$$

Is x small compared to 0.50; that is, is x/0.50 less than 0.05?

$$\frac{1.2 \times 10^{-2}}{0.500} = .024$$

Our assumption was valid.

What is $[H^+]$?

Look back at the concentration table:

$$[H^+] = 0.50 + x = 0.50 + 0.012 = 0.51\ M$$

$$pH = -\log(0.51) = 0.29$$

Now you're ready to begin and solve most acid-base equilibrium problems.

14-1. Given the equilibrium

$$HF(aq) + CH_3COO^-(aq) \rightleftharpoons F^-(aq) + CH_3COOH \qquad K_{eq} = 40$$

    a. Which two species in this equilibrium are Brönsted acids?

    b. Which two species in this equilibrium are Brönsted bases?

    c. Which is the stronger acid?

    d. Which is the stronger base?

14-2. Given: $HNO_2 + H_2O \rightleftharpoons H_3O^+ + NO_2^- \qquad K_a = 4.0 \times 10^{-4}$

    and $CH_3COOH + H_2O \rightleftharpoons H_3O^+ + CH_3COO^- \qquad K_a = 1.8 \times 10^{-5}$

    a. Which is the stronger acid, HNO_2 or CH_3COOH?

    b. Which is the stronger base, CH_3COO^- or NO_2^-?

14-3. Calculate the pH for a solution in which $[H^+] = 4.5 \times 10^{-6}$ M.

14-4. Calculate $[H^+]$ for a solution of pH = 3.47.

14-5. Calculate the pH of a 0.20 M CH_3COOH solution. $K_a = 1.8 \times 10^{-5}$

14-6. Calculate the pH of a 0.50 M NH_4Cl solution. $K_a = 5.6 \times 10^{-10}$

14-7. What percent of the acetic acid in 14-5 above has dissociated?

14-8. Calculate the pH of a 0.20 M NH_3 solution. $K_b = 1.8 \times 10^{-5}$

14-9. Calculate the pH of a 0.10 M HF solution, assuming that less than 5% of the HF dissociates. Is the assumption valid? $K_a = 7.2 \times 10^{-4}$

14-10. a. Calculate the pH of a 0.10 M H_2CO_3 solution. $K_{a_1} = 4.3 \times 10^{-7}$; $K_{a_2} = 5.6 \times 10^{-11}$

    b. Calculate the concentration of CO_3^{2-} in a 0.10 M H_2CO_3 solution.

CHAPTER FOURTEEN: SELF-TEST ANSWERS

14-1. a. HF and CH_3COOH

b. F^- and CH_3COO^-

c. HF

d. CH_3COO^-

14-2. a. HNO_2

b. CH_3COO^-

14-3. pH = 5.35

14-4. $[H^+] = 3.4 \times 10^{-4}\ M$

14-5. $\frac{[CH_3COO^-][H^+]}{[CH_3COOH]} = \frac{x^2}{0.20 - x} = 1.8 \times 10^{-5}$

$[H^+] = 1.9 \times 10^{-3}\ M$

pH = 2.72

14-6. $\frac{[NH_3][H^+]}{[NH_4^+]} = \frac{x^2}{0.50 - x} = 5.6 \times 10^{-10}$

$[H^+] = 1.7 \times 10^{-5}\ M$

pH = 4.77

14-7. $\frac{[CH_3COO^-]_{equil}}{[CH_3COOH]_{initial}} = \frac{1.9 \times 10^{-3}\ M}{2.0 \times 10^{-1}\ M} = 0.0095$; 0.95% dissociates

14-8. $\frac{[NH_4^+][OH^-]}{[NH_3]} = \frac{x^2}{0.20 - x} = 1.8 \times 10^{-5}$

$[OH^-] = 1.9 \times 10^{-3}\ M$

pOH = 2.72

pH = 11.28

14-9. $\frac{[H^+][F^-]}{[HF]} = \frac{x^2}{0.10 - x} = 7.2 \times 10^{-4} \approx \frac{x^2}{0.10}$

$x = 8.5 \times 10^{-3}$

$\frac{x}{0.10} = 0.085 = 8.5\%$; x is <u>not</u> negligible

14-10. Assume that H_2CO_3 is the principle source of $[H^+]$

$$\frac{[H^+][HCO_3^-]}{[H_2CO_3]} = \frac{x^2}{0.10 - x} = 4.3 \times 10^{-7}$$

$$[HCO_3^-] = [H^+] = x = 2.1 \times 10^{-4} \; M$$

Then: $$\frac{[H^+][CO_3^{2-}]}{[HCO_3^-]} = 5.6 \times 10^{-11}$$

$$[CO_3^{-2}] = 5.6 \times 10^{-11} \; M$$

CHAPTER FIFTEEN: AQUEOUS EQUILIBRIA

Here we meet more complex equilibria. We can continue to use the same approach we've used so far.

Problem 15-1:

What is the pH of a solution which is 0.150 M acetic acid and 0.200 M sodium acetate?

Solution:

All the species: CH_3COOH, CH_3COO^-, Na^+, H^+, OH^-, H_2O

The equilibria: $CH_3COOH \rightleftharpoons CH_3COO^- + H^+$ $K_a = 1.8 \times 10^{-5}$

$H_2O \rightleftharpoons H^+ + OH^-$ $K_w = 1.0 \times 10^{-14}$

Sodium ions are only "spectators," undergoing no reaction.

The principle equilibrium is the dissociation of acetic acid.

As acetic acid dissociates, the hydrogen ion concentration rises, suppressing the ionization of water. So, we'll assume that the water doesn't contribute a significant amount of hydrogen ion to the solution.

Set up the concentration table:

	$[CH_3COOH]$	$[CH_3COO^-]$	$[H^+]$
Initial	0.150	0.200	~0
Final	0.150 − x	0.200 + x	x

Use the equilibrium expression:

$$K_a = \frac{[CH_3COO^-][H^+]}{[CH_3COOH]} = \frac{(0.200 + x)(x)}{(0.150-x)} = 1.8 \times 10^{-5}$$

Let's assume x is small compared to 0.150 and to 0.200:

$$\left(\frac{0.200}{0.150}\right)x = 1.8 \times 10^{-5}$$

$$x = 1.35 \times 10^{-5}$$

$$[H^+] = 1.35 \times 10^{-5} \ M$$

$$pH = -\log(1.35 \times 10^{-5}) = 4.87$$

Remember to check the assumption: is x small compared to 0.150 and 0.200?

Problem 15-2:

What is the pH of a solution made by adding 10.0 mL of 0.100 M HCl to 100.0 mL of the solution from problem 15-1?

Solution:

We can solve this problem in a series of steps:

First: Let's calculate the concentrations of species in solution after dilution, before any reaction occurs:

The HCl: In 10.0 mL of 0.100 M HCl, there is:

$$10.0 \text{ mL} \times \frac{0.100 \text{ mmoles}}{\text{mL}} = 1.00 \text{ mmol HCl}$$

The final volume of the solution is 110.0 mL.

The concentration of HCl is then:

$$\frac{1.00 \text{ mmol HCl}}{110.0 \text{ mL}} = 9.09 \times 10^{-3} \, M$$

The CH_3COOH: In 100.0 mL of 0.150 M acetic acid, there are:

$$100.0 \text{ mL} \times \frac{0.150 \text{ moles}}{\text{liter}} = 15.0 \text{ mmol } CH_3COOH$$

15.0 mmol CH_3COOH is dissolved in 110.0 mL of solution.

The concentration of CH_3COOH is:

$$\frac{15.0 \text{ mmol } CH_3COOH}{110.0 \text{ mL}} = 0.136 \, M$$

Similarly, the concentration of sodium acetate is: 0.182 M.

HCl is a strong acid, which dissociates completely. Now we can set up our concentration table:

	$[CH_3COOH]$	$[CH_3COO^-]$	$[H^+]$
Initially, before any reaction occurs	0.136 M	0.182 M	$9.09 \times 10^{-3} \, M$

The hydrogen ion will react nearly completely with the acetate ion. [How do we know? What is the equilibrium constant for the reaction: $H^+(aq) + CH_3COO^-(aq) \rightleftharpoons CH_3COOH(aq)$?

The equilibrium constant for this reaction is:

$$K_{eq} = \frac{[CH_3COOH]}{[H^+][CH_3COO^-]} = \frac{1}{K_a} = 5.6 \times 10^4$$

This reaction goes nearly "to completion."]

Back to the concentration table:

	$[CH_3COOH]$	$[CH_3COO^-]$	$[H^+]$
Initially	0.136 M	0.182 M	9.09×10^{-3} M
After the reaction between H^+ and CH_3COO^-	(0.136 + .009) M	(0.182 − .009) M	~0

Now we have a mixture of acetic acid and acetate, very like the one in Problem 15-1: Back to the concentration table:

	$[CH_3COOH]$	$[CH_3COO^-]$	$[H^+]$
Initially	0.136 M	0.182 M	9.09×10^{-3} M
After the reaction between H^+ and CH_3COO^-	0.145	0.173	~0
At equilibrium	0.145 − x	0.173 + x	x

Then

$$K_a = \frac{[H^+][CH_3COO^-]}{[CH_3COOH]} = x\frac{(0.173 + x)}{(0.145 - x)} = 1.8 \times 10^{-5}$$

After the usual assumption, that x is small compared to 0.173 and 0.145:

$$\frac{x(0.173)}{(0.145)} = 1.8 \times 10^{-5}$$

$$[H^+] = x = 1.5 \times 10^{-5} \; M$$

$$pH = -\log [H^+] = 4.82$$

Problem 15-3:

What would the pH have been if 10.0 mL of HCl had been added to 100.0 mL of water, instead of 100.0 mL of the mixture in 15-1?

Solution:

From Problem 15-2: The HCl concentration would be:

$$9.09 \times 10^{-3} \; M$$

The HCl dissociates completely; therefore $[H^+] = 9.09 \times 10^{-3}$ M and the pH would be:

$$pH = -\log (9.09 \times 10^{-3}) = 2.05$$

From these three problems we can see that when 10.0 mL of 0.10 m HCl is added to 100.0 mL of water the pH changes from 7.00 to 2.05; when the HCl is added to 100.0 mL of the acetic acid–acetate mixture, the pH changes from 4.87 to 4.82. The mixture of the acid and its conjugate base is a "buffer" against changes in pH.

Buffers

A buffer is most effective when there are nearly equivalent concentrations of an acid and its conjugate base. A strong acid added to the buffer reacts with the conjugate base producing the weak acid. The change in the ratio of the base to its conjugate acid is small if the original concentrations of the conjugate acid and base are large. A strong base added to the buffer reacts with the conjugate acid producing the conjugate base. Again, the ratio of the conjugate base to its conjugate acid stays nearly constant.

We can rearrange the equilibrium expression

$$[H^+] = K_a \frac{[CH_3COOH]}{[CH_3COO^-]}$$

In our "best" buffer, we have equal concentrations of the acetic acid and acetate, then:

$$[H^+] = K_a$$

and pH = pK_a. If we want to buffer a solution at a particular pH, we should choose a conjugate acid–base mixture with a pK_a close to the pH.

Problem 15-4:

What concentrations of benzoic acid and sodium benzoate should be used to give a buffer of pH 4.00?

$$K_a \frac{[H^+][C_6H_5COO^-]}{[C_6H_5COOH]} = 6.1 \times 10^{-5}$$

Method I.

$$[H^+] = 10^{-pH} = 01 \times 010^{-4}$$

Then:
$$1.00 \times 10^{-4} \frac{[C_6H_5COO^-]}{[C_6H_5COOH]} = 6.1 \times 10^{-5}$$

$$\frac{[C_6H_5COO^-]}{[C_6H_5COOH]} = \frac{6.1 \times 10^{-5}}{1.0 \times 10^{-4}} = 0.61$$

We could use any concentrations of benzoate and benzoic acid, so long as the ratio is 0.61 to 1.00.

Method II.

If we take the logarithm of both sides of the equilibrium expression:

$$K_a = \frac{[H^+][C_6H_5COO^-]}{[C_6H_5COOH]} = 6.1 \times 10^{-5}$$

$$\log K_a = \log[H^+] + \log\frac{[C_6H_5COO^-]}{[C_6H_5COOH]} = -4.21_5$$

$$pK = pH - \log\frac{[C_6H_5COO^-]}{[C_6H_5COOH]} = 4.21_5$$

Rearranging: $pH = pK + \log\frac{[C_6H_5COO^-]}{[C_6H_5COOH]} = 4.21_5 + \log\frac{[C_6H_5COO^-]}{[C_6H_5COOH]}$

(This is the Henderson-Hasselbach equation, a different form of the equilibrium expression.)

Now: If we want a pH of 4.00:

$$4.00 = 4.21_5 + \log\frac{[C_6H_5COO^-]}{[C_6H_5COOH]}$$

Then:

$$\log\frac{[C_6H_5COO^-]}{[C_6H_5COOH]} = 4.00 - 4.21_5 = -0.21_5$$

$$\frac{[C_6H_5COO^-]}{[C_6H_5COOH]} = 10^{-.215} = .61$$

Titration Curves

You can calculate the pH at any point on the titration curve of an acid titrated with strong base. Each point is a problem of its own:

i) First, the acid alone

ii) Next, mixtures of the acid and conjugate-base (a buffer, if the acid is weak).

iii) At the equivalence point: the conjugate base alone.

iv) Beyond the equivalence point: solutions of the excess strong base.

The Titration of a Weak Acid with a Strong Base

First, calculate the initial pH from the concentration of the weak acid and its K_a.

Then, for each point before the equivalence point, do the stoichiometry problem first: let the hydroxide react completely with the weak acid. Then do the equilibrium problem for the resulting buffer.

At the equivalence point, do the equilibrium problem for the conjugate base alone.

The only problems you haven't met are the mixtures beyond the equivalence point.

Problem 15-5:

What is the pH of a solution after 60.00 mL of 0.100 M NaOH have been added to 50.00 mL of 0.100 M CH_3COOH? (This corresponds to the addition of 10.00 mL of NaOH past the equivalence point in the titration of 50.00 mL of 0.100 M CH_3COOH with 0.100 M NaOH.)

Solution:

First the stoichiometry problem:

We have

$$60.00 \text{ mL} \times \frac{0.100 \text{ moles NaOH}}{\text{liter}} = 6.00 \text{ mmoles NaOH}$$

$$50.00 \text{ mL} \times \frac{0.100 \text{ moles } CH_3COOH}{\text{liter}} = 5.00 \text{ mmoles } CH_3COOH$$

We have an excess of 1.00 mmoles of NaOH.

Let the NaOH react with the CH_3COOH:

	mmoles OH^-	mmoles CH_3COOH	mmoles CH_3COO^-
Initial	6.00	5.00	~0
Final	1.00	~0	5.00

Actually, the acetate ion will react with water to produce OH^-:

$$CH_3COO^-(aq) + H_2O(\ell) \rightleftharpoons CH_3COOH(aq) + OH^-(aq) \qquad K_b = 5.6 \times 10^{-10}$$

However, the excess OH^- in this solution suppresses this reaction. Let's assume that this reaction produces a negligible amount of OH^-. Then we can calculate the OH^- concentration from the excess OH^- in solution:

$$[OH^-] = \frac{1.00 \text{ mmole } OH^-}{110.0 \text{ mL}} = 9.1 \times 10^{-3} \ M$$

$$pOH = -\log (9.1 \times 10^{-3}) = 2.04$$

$$pH = 14.00 - pOH = 14.00 - 2.04 = 11.96$$

Acid-Base Indicators

An acid-base indicator changes color in the pH range from $pK_a - 1$ to $pK_a + 1$.

Problem 15-6:

What indicator should we use in the titration of 50.00 mL of CH_3COOH with 0.100 m NaOH?

Solution:

What is the pH at the end point?

First, the stoichiometry problem:

At the end point we've added 50.00 mL of 0.100 M NaOH to 50.00 mL of 0.100 M CH_3COOH:

$$mmol\ NaOH = 50.00\ mL \times 0.100\ M = 5.00\ mmol$$

$$mmol\ CH_3COOH = 50.00\ mL \times 0.100\ M = 5.00\ mmol$$

The strong base reacts completely with the acetic acid to form 5.00 mmol of the acetate ion.

Now, we can calculate the pH of the solution of acetate ion.

The concentration of acetate before it reacts with water is:

$$[CH_3COO^-] = \frac{5.00\ mmol}{100.0\ ml} = 5.00 \times 10^{-2}\ M$$

then equilibrium reaction is:

$$CH_3COO^-(aq) + H_2O(\ell) \rightleftharpoons CH_3COOH(aq) + OH^-(aq) \qquad K_b = 5.6 \times 10^{-10}$$

Set up the concentration table:

	$[CH_3COO^-]$	$[CH_3COOH]$	$[OH^-]$
Initially	$5.00 \times 10^{-2}\ M$	0	~0
Equilibrium	$5.00 \times 10^{-2}\ M - x$	x	x

Use the equilibrium expression:

$$K_b = \frac{[CH_3COOH][OH^-]}{[CH_3COO^-]} = \frac{x^2}{5.00 \times 10^{-2} - x} = 5.6 \times 10^{-10}$$

Assuming x is small compared to 5.00×10^{-2}:

$$\frac{x^2}{5.00 \times 10^{-2}} = 5.6 \times 10^{-10}$$

$$x^2 = 2.8 \times 10^{-11}$$

$$[OH] = x = 5.3 \times 10^{-6}$$

(*x* is indeed small compared to 5.00×10^{-2}.)

$$pOH = 5.28$$

$$pH = 14.00 - pOH = 14.00 - 5.28 = 8.72$$

Now, we need an indicator which changes color near this pH. Look at Figure 15.7. Thymol blue and phenolphthalein change near this pH.

Solubility Equilibria

For the reaction:

$$Ca(OH)_2(s) \rightleftharpoons Ca^{2+}(aq) = 2 \; OH^-(aq)$$

The equilibrium expression is:

$$K_{sp} = [Ca^{2+}][OH^-]^2 = 1.3 \times 10^{-6}$$

(Remember, pure solids and liquids don't appear in the equilibrium expression.)

Problem 15-7:

What is the solubility of calcium hydroxide?

Solution:

We can set up a concentration table:

	$[Ca^{2+}]$	$[OH^-]$
Initial	0	~0
At equilibrium	x	$2x$

And use the equilibrium expression:

$$K_{sp} = [Ca^{2+}][OH^-]^2 = x(2x)^2 = 1.3 \times 10^{-6}$$

$$4x^3 = 1.3 \times 10^{-6}$$

$$x^3 = 3.2_5 \times 10^{-7}$$

$$[Ca^{2+}] = x = 6.9 \times 10^{-3} \; M$$

$$\text{Note:} \quad [OH^-] = 2x = 1.4 \times 10^{-2} \; M$$

6.9×10^{-3} moles of $Ca(OH)_2$ can dissolve in a liter of water. The solubility of $Ca(OH)_2$ in water is 6.9×10^{-3} moles/liter.

Problem 15-8:

What is the solubility of $Ca(OH)_2$ in a 0.50 M NaOH solution?

Solution:

Set up a concentration table:

	$[Ca^{2+}]$	$[OH^-]$
Initial	0	0.50
At equilibrium	x	0.50 + 2x

Use the equilibrium expression:

$$K_{sp} = [Ca^{2+}][OH^-]^2 = x(0.50 + 2x)^2 = 1.3 \times 10^{-6}$$

Let's make the usual assumption, that 2x is small compared to 0.50:

$$x(.05)^2 = 1.3 \times 10^{-6}$$

$$[Ca^{2+}] = x = 5.2 \times 10^{-6} \ M$$

(Check the assumption.)

The solubility of $Ca(OH)_2$ in this solution is 5.2×10^{-6} moles/liter.

Problem 15-9:

What is the solubility of $Ca(OH)_2$ in a 0.50 M $CaCl_2$ solution?

Solution:

Set up the concentration table.

	$[Ca^{2+}]$	$[OH^-]$
Initial	0.50 M	~0
Final	0.50 + x	2x

And the equilibrium expression:

$$K_{sp} = [Ca^{2+}][OH^-]^2 = (0.50 + x)(2x)^2 = 1.3 \times 10^{-6}$$

Assume x is small compared to 0.50:

$$(0.50)(2x)^2 = 1.3 \times 10^{-6}$$

$$x = 6.5 \times 10^{-7} \ M$$

$$[Ca^{2+}] = 0.50 + 6.5 \times 10^{-7} = 0.50 \ M$$

$$[OH^-] = 2x = 1.3 \times 10^{-6} \ M$$

What is the solubility of $Ca(OH)_2$ in this solution? How many moles of $Ca(OH)_2$ would dissolve in a liter? x moles. The solubility, then, is 6.5×10^{-7} moles/liter.

As you can see, our general approach to equilibrium "conquers" solubility equilibria as well as acid–base equilibria.

One last solubility problem:

Problem 15-10:

When 60.0 mL of 0.0500 M NaOH is added to 40.00 mL of 0.0400 M $CaCl_2$, does a precipitate form?

$$K_{sp} = [Ca^{2+}][OH^-]^2 = 1.3 \times 10^{-6}$$

Solution:

We can use Q, the concentration expression.

If Q is greater than K_{sp}, a precipitate will form.

If Q is less than K_{sp}, no precipitate forms.

When the solutions are mixed together:

$$\text{mmol } Ca^{2+} = \frac{0.0400 \text{ mmoles}}{mL} \times 40.0 \text{ mL} = 1.60 \text{ mmol}$$

In the solution, before any reaction takes place:

$$[Ca^{2+}] = \frac{1.60 \text{ mmol}}{100.0 \text{ mL}} = 1.6 \times 10^{-2} \ M$$

$$\text{mmol } OH^- = \frac{0.0500 \text{ mmoles}}{mL} \times 60.0 \text{ mL} = 3.00 \text{ mmol}$$

In the solution, before any reaction takes place:

$$[OH^-] = \frac{3.00 \text{ mol}}{100.0 \text{ mL}} = 3.0 \times 10^{-2} \ M$$

$$Q = [Ca^{2+}][OH^-]^2 = (1.6 \times 10^{-2})(3.0 \times 10^{-2})^2 = 1.4 \times 10^{-5}$$

Q is larger than K: $Ca(OH)_2$ does precipitate.

Complex Ions

We can write stepwise formation constants for complexation equilibria. For example, silver ion and ammonia form a "diammine" complex. NH_3 in these reactions is bound to Ag^+; NH_3 is called the "ligand."

$$Ag^+(aq) + NH_3(aq) \rightleftharpoons Ag(NH_3)^+ \qquad K_1 = 2.1 \times 10^3$$

$$Ag(NH_3)^+(aq) + NH_3 \rightleftharpoons Ag(NH_3)_2^+ \qquad K_2 = 8.2 \times 10^3$$

Both equilibrium constants are large. We'll assume that, in excess NH_3, both of these reactions go to completion.

Problem 15-11:

What is the "free" (uncomplexed) silver ion concentration in a 0.050 M solution of $AgNO_3$ which is 1.50 M NH_3?

All the species: Ag^+, NO_3^-, NH_3, $Ag(NH_3)^+$, $Ag(NH_3)_2^+$, H^+, OH^-, H_2O

All the equilibria:

$$Ag^+(aq) + NH_3(aq) \rightleftharpoons Ag(NH_3)^+(aq) \qquad K_1 - 2.1 \times 10^3$$

$$Ag(NH_3)^+(aq) + NH_3 \rightleftharpoons Ag(NH_3)_2^+ \qquad K_2 = 8.2 \times 10^3$$

$$NH_3 + H_2O \rightleftharpoons NH_4^+ + OH^- \qquad K_b = 1.8 \times 10^{-5}$$

$$H_2O \rightleftharpoons H^+ + OH^- \qquad K_w = 1.0 \times 10^{-14}$$

Clearly the complexation equilibria are the principle equilibria. Let's assume that the first two reactions go nearly to completion. We can construct a concentration table, with initial concentrations, the concentrations after the first complexation step is complete and after the second complexation step is complete. Then we can let some of the $Ag(NH_3)_2^+$ dissociate to form $Ag(NH_3)^+$.

	$[Ag^+]$	$[Ag(NH_3)^+]$	$[Ag(NH_3)_2^+]$	$[NH_3]$
Initial	0.050 M	0	0	1.50 M
after the 1st step	0	0.050	0	1.50 − .050 = 1.45
after the 2nd step	0	0	0.050	1.45 − .050 = 1.40
At equilibrium	~0	x	0.050 − x	1.40 + x

Let's use K_2 to calculate the concentration of $Ag(NH_3)^+$:

$$K_2 = \frac{[Ag(NH_3)_2^+]}{[Ag(NH_3)^+][NH_3]} = \frac{0.050 - x}{x(1.40 + x)} = 8.2 \times 10^3$$

If x is small compared to 0.050:

$$\frac{0.050}{x(1.40)} = 8.2 \times 10^3$$

$$[Ag(NH_3)^+] = x = \frac{0.050}{(1.40)(8.2 \times 10^3)} = 4.4 \times 10^{-6} \ M$$

x is small enough to be ignored, when compared to 0.050 or 1.40.

Now, use K_1 to calculate $[Ag^+]$:

$$K_1 = \frac{[Ag(NH_3)^+]}{[Ag^+][NH_3]} = \frac{(4.4 \times 10^{-6})}{[Ag^+](1.40)} = 2.1 \times 10^3$$

$$[Ag^+] = \frac{(4.4 \times 10^{-6})}{(1.40)(2.1 \times 10^3)} = 1.5 \times 10^{-3} \ M$$

We've seen how the same approach can be used for all these equilibria: write all the species and all the equilibria; then find the principal equilibrium and construct a concentration table. Solve the problem, remembering to check your assumptions.

15-1   Calculate the pH of a buffer which is 0.200 M CH_3COOH and 0.300 M CH_3COO^-. $K_a = 1.8 \times 10^{-5}$

15-2.   Calculate the pH when 5.0 mmol of HCl is added to 100.0 mL of the buffer from 15-1.

15-3.   Calculate the pH when 25.0 mL of 0.100 M NaOH is added to 50.0 mL of 0.100 M CH_3COOH.

15-4.   Calculate the pH when 50.0 mL of 0.100 M NaOH is added to 50.0 mL of 0.100 M CH_3COOH.

15-5.   Calculate the pH when 60.0 mL of 0.100 M NaOH is added to 50.0 mL of 0.100 M CH_3COOH.

15-6.   How would you use CH_3COOH and $NaOOCCH_3$ to make a buffer of pH 5.00?

15-7.   At a certain temperature,

$$PbI_2(s) \rightleftharpoons Pb^{2+}(aq) + 2\ I^-(aq) \qquad K_{sp} = 1.4 \times 10^{-8}$$

Will a precipitate form when 50.0 mL of 1.0×10^{-2} M $Pb(NO_3)_2$ is added to 50.0 mL of 2.0×10^{-2} M NaI?

15-8.   How much PbI_2 will dissolve in 1.00 liter of water at this temperature?

15-9.   How much PbI_2 will dissolve in 1.00 liter of 0.10 M NaI at this temperature?

15-10. NH_3 is added to a 0.10 M solution of $AgNO_3$ until the total NH_3 concentration is 2.00 M.

$$Ag^+(aq) + NH_3(aq) \rightleftharpoons Ag(NH_3)^+(aq) \qquad K_1 = 2.1 \times 10^3$$

$$Ag(NH_3)^+(aq) + NH_3(aq) \rightleftharpoons Ag(NH_3)_2{}^+(aq) \qquad K_2 = 8.2 \times 10^3$$

a. What is the concentration of $Ag(NH_3)_2{}^+$ in this solution?

b. What is the concentration of uncomplexed NH_3 in this solution?

c. What is the concentration of $Ag(NH_3)^+$ in this solution?

CHAPTER FIFTEEN: SELF-TEST ANSWERS

15-1. $\frac{[x][0.300 + x]}{[0.200 - x]} \approx \frac{x[0.300]}{[0.200]} = 1.8 \times 10^{-5}$

$[H^+] = x = 1.2 \times 10^{-5}$

pH = +4.92

15-2.

Initially:	Finally:	At Equilibrium:
20.0 mmol CH_3COOH	25.0 mmol CH_3COOH	$[CH_3COOH] = 0.250\ M - x$
30.0 mmol CH_3COO^-	25.0 mmol CH_3COO^-	$[CH_3COO^-] = 0.250\ M + x$
5.0 mmol H^+	~0 mmol H^+ in 100.0 mL	$[H^+] = x$

$$\frac{[x][0.250 + x]}{[0.250 - x]} = 1.8 \times 10^{-5}$$

$[H^+] = x = 1.8 \times 10^{-5}$

pH = 4.74

15-3.

Initially:	Finally:	At Equilibrium:
2.50 mmol OH^-	0 mmol OH^-	$[CH_3COO^-] = 0.033\ M + x$
5.00 mmol CH_3COOH	2.50 mmol CH_3COO^-	$[CH_3COOH] = 0.033\ M - x$
	2.50 mmol CH_3COOH in 75.0 mL	$[H^+] = x$

$$\frac{[x][0.033 + x]}{[0.033 - x]} = 1.8 \times 10^{-5}$$

$[H^+] = 1.8 \times 10^{-5}$

pH = 4.74

15-4.

Initially:	Finally:	At Equilibrium:
	0 mmol OH^-	$[OH^-] = x$
5.00 mmol OH^-	0 mmol CH_3COOH	$[CH_3COOH] = x$
5.00 mmol CH_3COOH	5.00 mmol CH_3COO^- in 100.0 mL	$[CH_3COO^-] = 0.050 - x$

$$\frac{x^2}{0.050 - x} = K_b = 5.5 \times 10^{-10}$$

$[OH^-] = x = 5.2 \times 10^{-6}\ M$

pOH = 5.28

pH = 14.00 − 5.28 = 8.72

15-5.   Initially: Finally:

6.0 mmol OH⁻   1.0 mmol OH⁻

5.0 mmol CH₃COOH   5.0 mmol CH₃COO⁻
                        in 110 mL solution

The OH⁻ suppresses the reaction of CH₃COO⁻ with water:

Then $[OH^-] = \dfrac{1.0\ \text{mmol}}{110\ \text{mL}} = 9.1 \times 10^{-3}M$

pOH = 2.04

pH = 11.96

15-6. $\dfrac{[H^+][CH_3COO^-]}{[CH_3COOH]} = 1.8 \times 10^{-5}$

If $[H^+] = 1.0 \times 10^{-5}$: $\dfrac{[CH_3COO^-]}{[CH_3COOH]} = 1.8$

Use any mixture with the concentration of sodium acetate 1.8 times that of acetic acid.

15-7. $Q = [Pb^{2+}]_{actual}\ [I^-]^2_{actual}$

Initially 0.50 mmol Pb²⁺

and 1.00 mmol I⁻

in 100.0 mL of solution

$[Pb^{2+}]_{actual} = \dfrac{0.50\ \text{mmol}}{100.0\ \text{mL}} = 5.0 \times 10^{-3}\ M$

$[I^-]_{actual} = \dfrac{1.00\ \text{mmol}}{100.0\ \text{mL}} = 1.0 \times 10^{-2}\ M$

$Q = 5 \times 10^{-7}$

Q is greater than K_{eq}; precipitate forms

15-8. $[Pb^{2+}][I^-]^2 = 1.4 \times 10^{-8}$

$[Pb^{2+}] = x$

$[I^-] = 2x$

$(x)(2x)^2 = 1.4 \times 10^{-8}$

$[Pb^{2+}] = x = 1.5 \times 10^{-3} \ M$

15-9. $[Pb^{2+}][I^-]^2 = 1.4 \times 10^{-8}$

$[Pb^{2+}] = x$

$[I^-] = 0.10 + 2x$

$(x)(0.10 + 2x)^2 = 1.4 \times 10^{-8}$

$[Pb^{2+}] = x = 1.4 \times 10^{-6} \ M$

15-10. a. Assume both reactions go "to completion":

$[Ag(NH_3)_2{}^+] = 0.10 \ M$

b. $[NH_3] = 2.00 \ M - 2[Ag(NH_3)_2{}^+]$

$= 2.00 \ M - 2[0.10 \ M] = 1.80 \ M$

$$\frac{[Ag(NH_3)_2{}^+]}{[Ag(NH_3)^+][NH_3]} = 8.2 \times 10^3 = \frac{[0.10]}{[x][1.80 \ M]}$$

$[Ag(NH_3)^+] = x = 6.8 \times 10^{-6} \ M$

CHAPTER SIXTEEN: SPONTANEITY, ENTROPY AND FREE ENERGY

Once again, we take a look at thermodynamics.

The First Law

Energy is conserved, never created, nor destroyed. When energy is "released" by an exothermic reaction, potential energy stored in chemical bonds is converted to "microscopic" kinetic energy, the energy of motion of atoms and molecules. As the kinetic energy of the atoms and molecules in the system and the surroundings increases, the temperature rises. Sometimes we speak of this exothermic reaction as "giving off heat," although heat is not actually a substance. We can think of both "heat" and "work" as "methods of energy transfer" between the system and the surroundings.

Spontaneity

When "thermodynamicists" say that a reaction is spontaneous, they mean that the reaction will occur without any outside intervention. However, they leave to "kineticists" the question of how long the reaction will take.

You have some intuitive knowledge about the spontaneous direction of some processes. Consider, a ball which rolls down a hill and comes to a stop. The potential energy of the ball is first converted to kinetic energy of the ball, and then, by friction, into microscopic kinetic energy of the atoms and molecules in the ball and the hill and the air. We could imagine the ball's taking microscopic kinetic energy from the atoms and molecules in both the hill and the ball, and rolling back up the hill. We never see this happen, even though it wouldn't violate the First Law of Thermodynamics.

We never see "heat" flow from a cold object to a warmer one. We never see the molecules of a gas collect in one corner of a container. Again, neither of these processes is forbidden by the First Law of Thermodynamics. But, we never see these processes take place?

Why do these processes occur in only one direction?

The Second Law of Thermodynamics

Every spontaneous process leads to an increase in the entropy of the "universe." (Remember, the "universe" is the system plus the surroundings.)

Entropy

The entropy of a system is a quantitative measure of the "disorder" or "randomness" in a system. There are two kinds of "disorder" in a system: "positional disorder," depending on the arrangement of particles (molecules and atoms) in space and "thermal disorder," depending on how the total energy is distributed among all the particles. The more random the distribution of particles in space, the higher the entropy. The more random the distribution of energy among the particles, the higher the entropy. The entropy of a substance depends on the amount of the substance. The units of entropy are energy (Joules) per Kelvin.

Problem 16-1:

For each pair, choose the "system" with the higher entropy <u>and</u> explain your answer:

    a. 18.0 g of ice at 0°C   18.0 g of liquid water at 0°C

    b. 36.0 g of ice at 0°C   22.0 g of ice at 0°C

    c. A 10 liter container filled with   A 10 liter container filled with
        2 moles of NH_3 at 25°C   a mixture of 1 mole of N_2 and
                                         3 moles of H_2 at 25°C

    d. 18.0 g of liquid water at 25°C   18.0 g of liquid water at 75°C

Solution:

    a. The molecules in both the ice and liquid water have the same average kinetic energy. (Both samples are at the same temperature.) However, the water molecules in ice can only vibrate around their equilibrium positions, while the water molecules in liquid water can both vibrate and tumble. The entropy of 18.0 g of liquid water is higher than that of 18.0 g of ice.

    b. Entropy is an <u>intensive</u> thermodynamic property, that is, entropy is proportional to the amount of a substance. 36.0 g of ice at 0°C has the higher entropy of this pair.

    c. Each of these systems contains 2 moles of N atoms and 6 moles of H atoms in the same volume. However the container of NH_3 contains 2 moles of particles which can move independently, while the container of the mixture of N_2 and H_2 contains 4 moles of particles which can move independently. There is more positional "disorder" in the mixture of N_2 and H_2: the mixture has the higher entropy.

    d. The molecules of liquid water at 75°C have a higher average kinetic energy. There is more energy to be distributed among the molecules, and more ways to distribute it. As the temperature of a substance is raised, the entropy increases.

Exothermic Reactions

When a reaction produces "heat," the entropy of the surroundings increases. If the temperature of the surroundings is low, the added energy has a greater effect on the entropy than if the temperature of the surroundings is high at constant pressure and temperature.

$$\Delta S_{surr} = \frac{-\Delta H_{rxn}}{T}$$

Problem 16-2:

When 1.0 mole of steam condenses to form 1.0 mole of liquid water at 100°C, 40.7 kJ of "heat" is released. What is ΔS for the surroundings?

Solution:

$$\Delta S_{surr} = \frac{-\Delta H_{rxn}}{T} = \frac{-(-40.7 \text{ kJ})}{373 \text{ K}} = \frac{0.109 \text{ kJ}}{K} = \frac{109 \text{ J}}{K}$$

Remember, ΔS for the universe must increase for any spontaneous process and:

$$\Delta S_{univ} = \Delta S_{system} + \Delta S_{surr}$$

Free Energy

We can use our expression for ΔS_{surr}:

$$\Delta S_{univ} = \Delta S_{system} + \frac{-\Delta H_{system}}{T}$$

Now we can predict the spontaneity of a process by focussing only on the system.

We can define a new property, the Gibbs free energy:

$$G = H - TS$$

At constant temperature, the change in the Gibbs free energy for a system is:

$$\Delta G_{sys} = \Delta H_{sys} - T\Delta S_{sys}$$

If we divide both sides of this equation by $-T$:

$$\frac{-\Delta G_{sys}}{T} = \frac{-\Delta H_{sys}}{T} + \Delta S_{sys} = \Delta S_{univ}$$

For a spontaneous process, ΔS_{univ} must be positive. Therefore, for a spontaneous process at constant temperature and pressure, ΔG_{sys} must be negative.

Temperature and Spontaneity

Let's look at a table for a system, remembering:

$$\Delta G_{sys} = \Delta H_{sys} - T\Delta S_{sys}$$

If: and: then:

ΔH_{sys}	ΔS_{sys}	ΔG_{sys}	
+	−	+	The process is not spontaneous at any temperature.
−	+	−	The process is spontaneous at all temperatures.
+	+	?	The process will be spontaneous at high temperatures, not at low temperatures.
−	−	?	The process will be spontaneous at low temperatures, not at high temperatures.

Problem 16-3:

The reaction:

$$3\ Fe(s) + 2\ O_2(g) \rightarrow Fe_3O_4(s)$$

is enthalpy favored ($\Delta H^{\circ}_{rxn} = -1118$ kJ; the reaction is exothermic), but it is not entropy favored ($\Delta S^{\circ}_{rxn} = -346$ J/K; the entropy of the system decreases). Is this reaction spontaneous at 25°C? (The $\circ$ superscript indicates that a substance is present in its standard state.)

Solution:

Calculate ΔG°:

$$\Delta G^{\circ} = \Delta H^{\circ} - T\Delta S^{\circ}$$

$$= -1118 \text{ kJ} - 298 \text{ K } (-346 \text{ J/K})$$

$$= -1118 \text{ kJ} + 103 \text{ kJ}$$

$$= -1015 \text{ kJ}$$

ΔG° is negative: the reaction is spontaneous.

Why is ΔS° negative for the reaction:

$$3 \ Fe(s) + 2 \ O_2(g) \rightarrow Fe_3O_4(s)?$$

The product Fe_3O_4 is more ordered than the reactants; the product occupies a much smaller volume than the reactants. Therefore, the entropy of the product is less than the entropy of the reactants.

If the number of gas molecules <u>decreases</u> in the course of a reaction, ΔS_{rxn} will be <u>negative</u>. If the number of gas molecules <u>increases</u> in the course of a reaction, ΔS_{rxn} will be positive.

Entropies of Reaction

It is possible to calculate the absolute value of the entropy of a pure substance. We can use the absolute entropies of the products and reactants from a table to calculate the change in entropy for a reaction.

$$\Delta S°_{reaction} = \Sigma n S°_{products} - \Sigma n S°_{reactants}$$

or "the change in entropy for a reaction equals the sum of the absolute entropies for the reactants." Remember, absolute entropies are "per mole": The n's in the equation remind you to include the moles of both the reactants and the products in your calculation.

Problem 16-4:

Calculate ΔS° for the reaction:

$$2 \ H_2(g) + O_2(g) \rightarrow 2 \ H_2O(g)$$

Using these data:

substance	S° J/K mol
$H_2O(g)$	189
$H_2(g)$	131
$O_2(g)$	205

Solution:

$$\Delta S° = \Sigma n S°_{products} - \Sigma n S°_{reactants}$$

$$= 2 \cdot S°_{H_2O(g)} - 1 \cdot S°_{O_2(g)} - 2 \ S°_{H_2(g)}$$

$$= 2 \ mol \ (189) \ J/K \ mol) - 1 \ mol \ (205) \ J/K \ mol) - 2 \ mol \ (131 \ J/K \ mol)$$

$$= - 89 \ J/K$$

We can calculate $\Delta S°$, $\Delta H°$, and $\Delta G°$ for a reaction using the following equations:

$$\Delta S° = \Sigma n S°_{prod} - \Sigma n S°_{react}$$

$$\Delta H° = \Sigma n \Delta H°_f{}_{prod} - \Sigma n \Delta H°_f{}_{react}$$

and
$$\Delta G° = \Delta H° - T\Delta S°$$

or:
$$\Delta G° = \Sigma n \Delta G°_f{}_{prod} - \Sigma n \Delta G°_f{}_{react},$$

where $\Delta G°_f$ is the free energy of formation of a substance, the change in free energy for the formation of one mole of a substance (in its standard state) from its constituent elements (in their standard states).

Problem 16-5:

Calculate $\Delta G°$ for the reaction at 25°C and 1 atm:

$$2\ H_2(g) + O_2(g) \rightarrow 2\ H_2O(g)$$

From the Data:

Substance	$\Delta G°_f$
$H_2(g)$	0 kJ/mol
$H_2O(g)$	-289
$O_2(g)$	0

Solution:

By definition, $\Delta G°_f$ for an element is zero.

$$\Delta G° = 2\ \Delta G°_{f_{H_2O(g)}} - \Delta G°_{f_{O_2(g)}} - 2\ \Delta G°_{f_{H_2(g)}}$$

$$= 2\ mol\ (-289\ kJ/mol) = -578\ kJ$$

Dependence of Free Energy on Pressure

The entropy of one mole of a gas in a 20.0 liter container is larger than the entropy of one mole of the same gas in a 10.0 liter container: there are more ways for the molecules of the gas to be arranged in space in the larger volume.

If the entropy of a gas depends on the volume (and the pressure), then the free energy of the gas must also depend on the volume (and the pressure):

$$G = G° + RT\ \ln p$$

where $G°$ is the free energy of the gas at 1 atmosphere pressure and G is the free energy of the gas at pressure p.

How does the free energy change for a <u>reaction</u> depend on the pressure?

Methane "burns" in oxygen to form carbon dioxide and water:

$$CH_4(g) + 2\ O_2(g) \rightarrow CO_2(g) + 2\ H_2O(g)$$

For this reaction:

$$\Delta G = G_{CO_2} + 2\ G_{H_2O} - G_{CH_4} - 2\ G_{O_2}$$

If we use $G = G° + RT \ln p$ for each component:

$$\Delta G = G°_{CO_2} + RT \ln p_{CO_2} + 2(G°_{H_2O} + RT \ln p_{H_2O})$$
$$- (G°_{CH_4} + RT \ln p_{CH_4}) - 2(G°_{O_2} + RT \ln p_{O_2})$$

Collect all the G° terms and all the RT ln p terms:

$$\Delta G = G°_{CO_2} + 2\ G°_{H_2O} - G°_{CH_4} - 2\ G°_{O_2}$$
$$+ RT(\ln p_{CO_2} + 2 \ln p_{H_2O} - \ln p_{CH_4} - 2 \ln p_{O_2})$$

$$\Delta G° = G°_{CO_2} + 2\ G°_{H_2O} - G°_{CH_4} - 2\ G°_{O_2}$$

and $\ 2 \ln p_{H_2O} = \ln p_{H_2O}^2 \ $ and $\ 2 \ln p_{O_2} = \ln p_{O_2}^2$

Then we have:

$$\Delta G = \Delta G° + RT\ (\ln p_{CO_2} + \ln p_{H_2O}^2 - \ln p_{CH_4} - \ln p_{O_2}^2)$$

Combining logarithms [remember $\ln a + \ln b = \ln(a \cdot b)$

and $\ln c - \ln d = \ln(c/d)$]

We have:

$$\Delta G = \Delta G° + RT \ln \left[\frac{p_{CO_2} \cdot p_{H_2O}^2}{p_{CH_4} \cdot p_{O_2}^2}\right]$$

The expression in square brackets is the reaction quotient, Q, from the equilibrium chapter.

You don't need to be able to derive this expression. Just remember that, for any reaction:

$$\Delta G = \Delta G° + RT \ln Q$$

You can calculate $\Delta G°$ for a reaction from tabulated values of free energies of formation of reactants and products. Using $\Delta G°$ and Q you can decide if the reaction mixture is at equilibrium. (At equilibrium, $\Delta G = 0$.)

Problem 16-6:

For the reaction:

$$2\ NO_2(g) \rightarrow N_2O_4(g)$$

$$\Delta G° = -5.39\ kJ\ at\ 25°C$$

If a mixture contains NO_2 with a partial pressure of 0.20 atm and N_2O_4 with a partial pressure of 0.10 atm, is the gas mixture at equilibrium? If not, in which direction will the reaction take place?

Solution:

$$\Delta G = \Delta G° + RT\ ln\ Q$$

$$= \Delta G° + \frac{8.3148\ J}{K\ mol}\ (298\ K)\ ln\ \frac{P_{N_2O_4}}{(P_{NO_2})^2}$$

$$= -5.39\ kJ + \frac{8.3148\ J}{K\ mol}\ (298\ K)\ ln\ \left[\frac{.10}{(20)^2}\right]$$

$$= -5.39\ kJ + 2.48 \times 10^3\ J\ ln\ 2.5$$

$$= -5.39 \times 10^3\ J + 2.48 \times 10^3\ J\ (.916)$$

$$= -5.39 \times 10^3\ J + 2.27 \times 10^3\ J = -3.12 \times 10^3\ J$$

As ΔG isn't equal to zero, the gas mixture is not at equilibrium. Since ΔG is negative, the reaction will proceed in the forward direction.

Problem 16-7:

Consider the reaction:

$$2\ NO_2(g) \rightarrow N_2O_4(g)$$

For a mixture which contains NO_2 with a partial pressure of 0.10 atm and N_2O_4 with a partial pressure of 0.20 atm, ΔG for this reaction is found to be $+2.04 \times 10^3$ J. In which direction will the reaction proceed?

Solution:

ΔG is greater than zero. The reaction can't proceed forward. ΔG for the reverse reaction will be negative: the reaction occurs in the reverse direction.

Free Energy and the Equilibrium Constant

For any reaction mixture:

$$\Delta G = \Delta G^\circ + RT \ln Q$$

At equilibrium: $\Delta G = 0$. (The mixture has reached a minimum free energy) and $Q = K$, the equilibrium constant.

Then, at equilibrium:

$$\Delta G = \Delta G^\circ + RT \ln Q$$

becomes:

$$0 = \Delta G^\circ + RT \ln K$$

and

$$\Delta G^\circ = -RT \ln K$$

We can use the tabulated values of ΔG_f° to calculate ΔG° for a reaction and then calculate K.

Problem 16-8:

Calculate the equilibrium constant for the reaction:

$$2\ NO_2(g) \rightarrow N_2O_4(g) \quad \text{at } 25^\circ C$$

Given the following data:

Substance	ΔG_f°	kJ/mol
NO_2	51.84	
N_2O_4	98.29	

Solution:

First calculate ΔG°:

$$\Delta G^\circ_{reaction} = \Sigma n \Delta G^\circ_{f\ products} - \Sigma n \Delta G^\circ_{f\ reactants}$$

$$= 1 \cdot \Delta G^\circ_{f\ N_2O_4} - 2 \cdot \Delta G^\circ_{f\ NO_2}$$

$$= 98.29\ kJ - 2(51.84)\ kJ$$

$$= -5.39\ kJ$$

Then calculate K_p:

$$\Delta G^\circ = -RT \ln K_p$$

$$-5.39 \text{ kJ} = \frac{-8.3148 \text{ J}}{K} \text{ (298 K) } \ln K_p$$

$$-5.39 \times 10^3 \text{ J} = -2.48 \times 10^3 \text{ J } \ln K_p$$

$$\ln K_p = \frac{-5.39 \times 10^3 \text{ J}}{-2.48 \times 10^3 \text{ J}} = 2.17$$

$$K_p = e^{2.17}$$

Use your e^x key, or your inverse key and ln key:

$$K_p = 8.8$$

ΔG° is negative. Therefore, K_p must be positive. This agrees with our result.

Let's return to Problem 16-6 using K_p instead of ΔG°:

Problem 16-9:

For the reaction:

$$2 \text{ NO}_2(g) \rightarrow \text{N}_2\text{O}_4(g) \quad \text{at } 25^\circ, \quad K_p = 8.8$$

"If a gas mixture contains NO_2 with a partial pressure of 0.20 atm and N_2O_4 with a partial pressure of 0.10 atm, is the gas mixture at equilibrium?

Solution:

Use the reaction quotient Q:

$$Q = \frac{P_{N_2O_4}}{P_{NO_2}^2} = \frac{0.10}{(0.20)^2} = 2.5$$

Q is not equal to K: therefore the system is not at equilibrium.

Q is less than K: therefore, the reaction will occur in the forward direction to produce more N_2O_4 and reduce the amount of NO_2.

We can use ΔG to determine whether a reaction mixture is at equilibrium, and, if it is not, the direction of the spontaneous process. ΔG also gives us the maximum possible useful work we can obtain from a process (at constant temperature and pressure). In any real process, some of the energy produced is "lost" to the surroundings as microscopic kinetic energy (thermal energy).

The energy isn't actually lost, but its usefulness is.

CHAPTER SIXTEEN: SELF-TEST

16-1. When 1.0 mole of ice melts to form 1.0 mole of liquid water at 0.0°C, 6.0 kJ of heat is "removed" from the surroundings. What is ΔS for the surroundings?

16-2. Complete this table: Decide if the process is spontaneous at

i) all temperatures

ii) no temperatures

iii) high temperatures

or iv) low temperatures

	ΔS_{rxn}	ΔH_{rxn}	ΔG_{rxn}
a.	−	+	_____
b.	−	−	_____
c.	+	+	_____
d.	+	−	_____

16-3. For the reaction:

$$Ag_2O(s) + H_2(g) \rightarrow 2\ Ag(s) + H_2O(g)$$

$\Delta H^\circ_{298} = -210.7$ kJ

and $\Delta S^\circ_{298} = +22.4$ J/K

Is this reaction spontaneous at 298 K?

16-4. Calculate ΔS° for the reaction:

$$N_2O_4(g) \rightleftharpoons 2\ NO_2(g)$$

Using these data:

Substance	S° J/K mol
$N_2O_4(g)$	304.3
$NO_2(g)$	210.6

16-5. Calculate $\Delta G°$ for the reaction: $CH_4(g) + 2\ O_2(g) \rightarrow CO_2(g) + 2\ H_2O(g)$ at 298 K (25°C).

$$\Delta G_f^\circ$$

$CH_4(g)$	−51 kJ/mol
$CO_2(g)$	−394 kJ/mol
$H_2O(g)$	−289 kJ/mol
$O_2(g)$	0 kJ/mol

16-6. For the reaction $N_2O_4(g) \rightarrow 2\ NO_2(g)$

$\Delta G° = +5.39$ kJ at 25°C.

What is K_p for this reaction?

16-7. In a mixture of N_2O_4 and NO_2 at 28°C, the partial pressure of N_2O_4 is 0.20 atm and the partial pressure of NO_2 is 0.10 atm.

a. (Optional) Calculate ΔG for the reaction in this mixture.

b. Will the reaction occur in the forward direction to approach equilibrium? (Is ΔG positive or negative?)

c. How much work can the system do as it approaches equilibrium?

CHAPTER SIXTEEN: SELF-TEST ANSWERS

16-1. $\Delta S_{surr} = \dfrac{-\Delta H}{T} = \dfrac{-(6.0 \text{ kJ})}{273} = -22 \text{ J/K}$

16-2.

	ΔG_{rxn}	Spontaneous
a.	+	No
b.	?	At low temperatures
c.	?	At high temperatures
d.	–	Yes

16-3. $\Delta G° = \Delta H° - T\Delta S°$

$= -210.7 \text{ kJ} - (298 \text{ K})(22.4 \text{ J/K})$

$= -210.7 \text{ kJ} - 6.68 \text{ kJ}$

$= -217.4 \text{ kJ}$

The reaction is spontaneous

16-4. $\Delta S° = 2 S°_{NO_2} - S°_{N_2O_4}$

$= 2(210.6) - 304.3 = 116.9 \text{ JK}^{-1}$

16-5. $\Delta G° = 2 \Delta G°_{f(H_2O)} + \Delta G°_{f(CO_2)} - 2 \Delta G°_{f(O_2)} - \Delta G°_{f(CH_4)}$

$= 2(-289 \text{ kJ}) + (-394 \text{ kJ}) - 2(0 \text{ kJ}) - (-51 \text{ kJ})$

$= -921 \text{ kJ}$

16-6. $\Delta G° = 5.39 \times 10^3 \text{ J} = - RT \ln K_p$

$\ln K_p = \dfrac{-5.39 \times 10^3 \text{ J}}{8.3148 \text{ J/K mol } (298 \text{ K})}$

$= -2.18$

$K_p = e^{-2.18} = 1.1 \times 10^{-1}$

16-7. a. (Optional) $\Delta G = \Delta G° + Rt \ln Q = \Delta G° + RT \ln \dfrac{(P_{NO_2})^2}{(P_{N_2O_4})}$

$= 5.39 \text{ kJ} + \dfrac{8.3148 \text{ J}}{\text{K mol}} (298) \ln \dfrac{(0.10)^2}{(0.20)}$

$= 5.39 \text{ kJ} + 2.48 \text{ kJ} \ln (0.05)$

$= 5.39 \text{ kJ} - 7.42 \text{ kJ} = -2.03 \text{ kJ}$

b. ΔG is negative; the reaction will proceed in the forward direction.

c. The system can do up to 2.03 kJ of work as it approaches equilibrium.

CHAPTER SEVENTEEN: ELECTROCHEMISTRY

Remember, from Chapter 4, that a "redox reaction" involves reduction, the gain of electrons, and oxidation, the loss of electrons. Now we wish to harness a redox reaction to produce an electrical current, or to use an electrical current to carry out a redox reaction.

Galvanic Cells

A galvanic cell is a "battery," in which a spontaneous chemical reaction produces an electrical current.

In order to harness a redox reaction to do electrical work we need to separate the oxidation half-reaction from the reduction half-reaction in two separate compartments. We need a connection between the two compartments to allow ions to flow without allowing the solutions to mix. The connection can be a "salt-bridge" (a u-shaped tube filled with agar and a salt solution) or a porous disk between the two solutions. (See Figures 17.2a and b in the text.)

Let us construct a galvanic cell like the one in 17.2b, with a zinc rod dipping into a solution of 1 M zinc nitrate in the left hand compartment, and a copper rod dipping into a solution of 1 M copper nitrate in the right hand compartment, with the two solutions separated by a porous disk.

If we use a "voltmeter" to connect the two rods (the electrodes) we find the voltage difference between the two rods to be about 1.1 volts, with electrons flowing from the zinc electrode to the copper electrode. As negatively charged electrons flow through the external circuit towards the copper electrode, the copper electrode must be more positive than the zinc electrode.

We can use a shorthand notation to describe this electrochemical cell:

$$^- \ Zn(s) \ | \ Zn(NO_3)_2(aq) \ || \ Cu(NO_3)_2(aq) \ | \ Cu(s) \ ^+$$

where the single vertical line represents a phase boundary and the double vertical lines represent the porous disk or salt bridge. The signs of the electrodes must be determined experimentally.

Cell Reactions

The convention for writing the cell reaction for a cell is:

- Write the half-reaction for the left electrode as oxidation.

- Write the half-reaction for the right electrode as reduction.

$$Zn(s) \rightarrow Zn^{2+}(aq) + 2e^-$$

$$2e^- + Cu^{2+}(aq) \rightarrow Cu(s)$$

Then, to write the cell reaction, we combine these two half-reactions:

$$Zn(s) + Cu^{2+}(aq) \rightarrow Zn^{2+}(aq) + Cu(s)$$

Oxidation is occurring at the left hand electrode: The electrode where oxidation occurs is the anode. Reduction occurs at the cathode, in this cell, the right hand electrode.

Electrode Potentials

We can think of the cell reaction in this cell as arising from a competition between the zinc and copper ions for electrons:

$$Zn^{2+}(aq) + 2e^- \rightarrow Zn(s)$$

and $\quad Cu^{2+}(aq) + 2e^- \rightarrow Cu(s)$

or $\quad Zn^{2+} \ldots 2e^- \ldots Cu^{2+}$

Which ion has the greater relative "electron-pulling-power"? Clearly Cu^{2+} does, as Cu^{2+} is reduced in this cell, while Zn metal is oxidized. We can use the measured cell voltage to determine relative "electron-pulling-power": Cu^{2+} has approximately 1.1 volts more "electron-pulling-power" than Zn^{2+}.

We can't measure the voltage (potential) of a single electrode; we can only measure voltage differences between two electrodes. We can make a table of relative single electrode voltages (potentials), if we arbitrarily choose one electrode to have a zero voltage (potential). Chemists have chosen the potential for the half-reaction:

$$2 H^+(aq, 1.0 \, M) + 2e^- \rightarrow H_2(g, 1 \text{ atm})$$

to be defined as exactly 0.00 volts.

Then we can assign the voltage for this half-reaction:

$$Zn^{2+}(aq) + 2e^- \rightarrow Zn(s)$$

by measuring the voltage for the cell pictured in 17.5b:

$$^- Zn \mid Zn^{2+}(aq, 1.0 \, M) \mid\mid H^+(aq, 1.0 \, M) \mid H_2(g, 1 \text{ atm}) \mid Pt \, ^+$$

We find the voltage to be 0.76 volts, with electrons flowing from the zinc electrode to the platinum electrode. H^+ then has a greater "electron-pulling-power" than Zn^{2+}. The reduction half-reaction: $2 H^+(aq) + 2e^- \rightarrow H_2(g)$ has a greater "tendency-to-occur" than $Zn^{2+}(aq) + 2e^- \rightarrow Zn(s)$. We can assign relative voltages to the half-reactions:

$$2 H^+(aq) + 2e^- \rightarrow H_2(g) \qquad 0.00 \text{ V}$$

$$Zn^{2+}(aq) + 2e^- \rightarrow Zn(s) \qquad -0.76 \text{ V}$$

258

If we measure the voltage for the cell:

$$+ \; Cu \; | \; Cu^{2+}(aq, \; 1.0 \; M) \; || \; H^+(aq, \; 1.0 \; M) \; | \; H_2(g, \; 1 \; atm) \; | \; Pt \; -$$

we find the voltage to be 0.34 volts, with electrons flowing from the platinum electrode to the copper electrode: Cu^{2+} has a greater "electron-pulling-power" than H^+. Cu^{2+} is more easily reduced than H^+. The relative voltage for this half-reaction is then:

$$Cu^{2+}(aq) + 2e^- \rightarrow Cu(s) \quad +0.34 \; volts$$

Table 17.1 lists the standard (all species present in their standard states at 1 atm pressure) reduction potentials, $E°$, at 298 K for many half-reactions in acidic solution.

The more positive the reduction potential for a half-reaction is, the greater the tendency for the reduction to take place, and the stronger the oxidizing agent is. From these reduction potentials, Cu^{2+} is a stronger oxidizing agent than H^+ which is a stronger oxidizing agent than Zn^{2+}; or Cu^{2+} is more easily reduced than H^+ which is more easily reduced than Zn^{2+}.

Problem 17-1:

Use the data in Table 17.1 to choose the stronger oxidizing agent in each pair:

    i) Sn^{2+} or Ag^+

    ii) Br_2 or Cl_2

    iii) Zn^{2+} or Al^{3+}

Solution:

The stronger oxidizing agent is most easily reduced.

    i) $Sn^{2+} + 2e^- \rightarrow Sn \quad -0.14 \; V$

         $Ag^+ + e^- \rightarrow Ag \quad +0.80 \; V$

         Ag^+ is the stronger oxidizing agent: the reduction potential is more positive.

    ii) $Br_2 + 2e^- \rightarrow 2 \; Br^- \quad +1.09 \; V$

         $Cl_2 + 2e^- \rightarrow 2 \; Cl^- \quad +1.36 \; V$

         Cl_2 is the stronger oxidizing agent.

    iii) $Zn^{2+} + 2e^- \rightarrow Zn \quad -0.76 \; V$

         $Al^{3+} + 3e^- \rightarrow Al \quad -1.71 \; V$

         Zn^{2+} is the stronger oxidizing agent.

Calculating Cell Voltages

We can use the standard reduction potentials to calculate the standard cell voltage for a cell.

For example, consider the cell:

$$Cu(s) \mid Cu^{2+}(aq, 1.0\ M) \parallel Fe^{3+}(aq, 1.0\ M), Fe^{2+}(aq, 1.0\ M) \mid Pt$$

What voltage will this cell develop? In which direction do electrons flow outside the cell?

First, write the two <u>reduction</u> half-reactions, and look up the reduction potentials:

$$Fe^{3+} + e^- \rightarrow Fe^{2+} \qquad E° = 0.77\ V$$

$$Cu^{2+} + 2e^- \rightarrow Cu(s) \qquad E° = 0.34\ V$$

Fe^{3+} is a stronger oxidizing agent than Cu^{2+}. Cu will be oxidized, Fe^{3+} will be reduced. We can combine these two half-reactions to give the spontaneous cell reaction:

$$2(Fe^{3+} + e^- \rightarrow Fe^{2+})$$

$$Cu(s) \rightarrow Cu^{2+} + 2e^-$$

$$\overline{\phantom{2 Fe^{3+} + Cu(s) \rightarrow 2 Fe^{2+} + Cu^{2+}}}$$

$$2\ Fe^{3+} + Cu(s) \rightarrow 2\ Fe^{2+} + Cu^{2+}$$

Electrons flow from the copper electrode to the platinum electrode; the platinum electrode is positive; oxidation occurs at the copper electrode. There are two methods for calculating E° for the reaction:

Method I

$$E°_{cell} = E°_{reduction} + E°_{oxidation}$$
$$\text{half-reaction} \qquad \text{half-reaction}$$

Where E° for an oxidation is −E° for the reverse of the oxidation reaction, the corresponding reduction half-reaction. In this case:

$$2(Fe^{3+} + e^- \rightarrow Fe^{2+}) \qquad E° = +0.77\ V$$

$$Cu(s) \rightarrow Cu^{2+} + 2e^- \qquad E° = -0.34\ V$$

(Where, for $Cu^{2+} + 2e^- \rightarrow Cu(s)$, $E° = +0.34\ V$.)

Then, for the cell reaction, $E° = 0.77\ V + (-0.34\ V) = 0.43\ V$

Method II

$$E^\circ_{cell} = E^\circ_{reduction\ pot'l} - E^\circ_{reduction\ pot'l}$$
$$\text{right electrode} \quad\quad \text{left electrode}$$

In this case: $E^\circ_{cell} = 0.77\ V - (0.34\ V) = 0.43\ V$

Using this convention, we always write the cell reaction with oxidation at the left electrode and reduction at the right. If E°_{cell} is positive, the cell reaction is spontaneous; if E°_{cell} is negative, the reverse of the cell reaction is spontaneous.

E° is the voltage the cell will develop if all the reactants and products are present in their standard states.

Notice that, even though we may multiply half-reactions by integers to combine them to give an overall reaction with no electrons in the balanced equation, we don't multiply the half-reaction potentials: the E°'s are on a per-mole-of-electrons basis, and can be combined without multiplying.

Problem 17-2:

Calculate the voltage developed by the cell:

$$Ag(s)\ |\ Ag^+(aq,\ 1.0\ M)\ ||\ Cu^{2+}(aq,\ 1.0\ M)\ |\ Cu(s)$$

Write the cell reaction, determine in which direction electrons will flow in the external circuit, and label each electrode as + or −.

Solution:

Method I

Write the reduction half-reactions

$$Ag^+ + e^- \rightarrow Ag \quad\quad +0.80\ V$$

$$Cu^{2+} + 2e^- \rightarrow Cu \quad\quad +0.34\ V$$

Ag^+ is a stronger oxidizing agent than Cu^{2+}; that is, Ag^+ can oxidize Cu. The spontaneous reaction will be:

$$2(Ag^+ + e^- \rightarrow Ag) \quad\quad +0.80\ V$$

$$Cu \rightarrow Cu^{2+} + 2e^- \quad\quad -0.34\ V$$

$$\overline{}$$

$$2\ Ag^+ + Cu \rightarrow 2\ Ag + Cu^{2+}$$

$$E^\circ = E^\circ_{Ag^+/Ag\ (reduction)} + E^\circ_{Cu/Cu^{2+}\ (oxidation)} = 0.80\ V + (-0.34\ V) = 0.46\ V$$

Electrons will travel from the copper electrode to the silver electrode: The silver electrode is more positive than the copper electrode:

$$+\ Ag(s)\ |\ Ag^+(aq\ 1.0\ M)\ ||\ Cu^{2+}(aq,\ 1.0\ M)\ |\ Cu(s)\ -$$

261

Method II

$$Ag(s) \mid Ag^+(aq, 1.0 \ M) \mid\mid Cu^{2+}(aq, 1.0 \ M) \mid Cu(s)$$

Write the cell reaction, with oxidation at the left electrode, and reduction at the right electrode:

$$2(Ag \rightarrow Ag^+ + e^-)$$

$$Cu^{2+} + 2e^- \rightarrow Cu$$

$$\overline{\hspace{4cm}}$$

$$2\ Ag + Cu^{2+} \rightarrow 2\ Ag^+ + Cu$$

Calculate $E^\circ_{cell} = E^\circ_{reduction, \ right} - E^\circ_{reduction, \ left}$

$$= 0.34 \ V \qquad\qquad -0.80 \ V$$

$$= -0.46 \ V$$

E°_{cell} is negative: therefore the reverse of the cell reaction occurs spontaneously: electrons flow from the copper electrode to the silver electrode, etc. The silver electrode is the positive electrode; the copper electrode is the negative electrode.

Method I combines the reduction potential for the cathode with the oxidation potential for the anode. Method II looks at the difference between reduction potentials for both electrodes. Each method leads to the same answer.

Current and Charge

Electric current is the rate-of-flow-of-electric-charge; current is measured in amperes (or amps), where one ampere equals one Coulomb of charge transferred per second. Then, we can calculate the charge transferred:

$$q = i \times t$$

where q is the charge transferred, in Coulombs

   i is the current, in amperes (c/s)

 and t is the time, in seconds.

The charge on an electron is $1.60219 \times 10^{-19} C$. The charge corresponding to one mole of electrons, the Faraday, is then:

$$F = 1.60219 \times 10^{-19} \ C/electron \times 6.02205 \times 10^{23} \ electrons/mole$$

$$= 96,485 \ C/mole$$

We can calculate the _moles_ of electrons transferred:

$$q = nF \text{ or } n = \frac{q}{F}$$

where q is the charge transferred, in Coulombs

F is the Faraday constant (96,485 C/mole)

and n is the number of moles of electrons.

Problem 17-3:

A 2.50 amp current is passed through a solution of copper (II) ion for 3 hours and 45 minutes. How much copper is deposited on the cathode?

Plan:

current → charge → moles of electrons → moles of copper → mass of copper

Solution:

First, calculate the time in seconds:

3 hours × 60 min/hour × 60 seconds/min = 10,800 s

+ 45 min × 60 seconds/min = 2,700 s

                            13,500 s

Then, calculate the charge:

$$q - i \cdot t$$

$$= 2.50 \text{ amps} \times 1.35 \times 10^4 \text{ s}$$

$$= 2.50 \text{ C/s} \times 1.35 \times 10^4 \text{ s} = 3.37_5 \times 10^4 \text{ C}$$

Next calculate moles of electrons:

$$n = \frac{q}{F}$$

$$= \frac{3.37_5 \times 10^4 C}{96,485 \text{ C/mole e}^-} = 0.350 \text{ moles e}^-$$

Then, calculate the moles of copper from the half-reaction:

$$Cu^{2+} + 2e^- \rightarrow Cu$$

$$\text{moles of Cu deposited} = \frac{\text{moles of electrons passed through the solution}}{2}$$

$$\text{or moles of Cu} = \text{moles of electrons} \times \frac{1 \text{ mole Cu}}{2 \text{ moles e}^-}$$

$$\text{moles of Cu} = \frac{0.350}{2} = 0.175 \text{ moles}$$

Finally, calculate the mass of copper deposited:

$$\text{grams of Cu} = 0.175 \text{ moles} \times 63.55 \text{ g/mole}$$

$$= 11.1 \text{ g Cu}$$

Electrical Work

A ball rolls down a hill, from a higher potential energy to a lower potential energy. This change in potential energy can be harnessed to do work. However, some of the potential energy will be "wasted," lost as increased microscopic kinetic energy ("heat") of the molecules in the ball, the hill and the surrounding air.

Similarly, current will "flow" from a point of higher potential in a circuit to a point of lower potential. We can harness this current flow to do work, but, again, some energy will be "wasted." The actual work we obtain is always less than the maximum possible work.

The magnitude of the work:

$$| \text{work (Joules)} | = \text{potential (volts)} \times \text{charge (Coulombs) transferred}$$

When one Coulomb of charge is transferred from one point in a circuit to another point at one volt potential lower, the current can do one Joule of work. On the other hand, it takes one Joule of work to move one Coulomb of charge from a point in a circuit to another point at a potential one volt higher.

If we use our electrochemical cell to do work on the surroundings: E_{cell} is positive, and w (work done on the system) is negative. Then:

$$w\text{(Joules)} = -E_{cell}\text{(Volts)} \cdot q\text{(Coulombs)}$$

where E_{cell} is the actual voltage while current is flowing. The maximum possible work is:

$$w_{max} = -E_{cell_{max}} \cdot q$$

If n moles of electrons are transferred:

$$w_{max} = -nFE_{max}$$

For a process at constant pressure: the maximum useful work equals the change in free energy for the process:

$$\Delta G = w_{max}$$

Then, for a galvanic cell, the change in free energy is given by:

$$\Delta G = w_{max} = -nFE_{max}$$

264

From now on, we'll use E to indicate the maximum potential developed by a cell. If E_{cell} is positive, ΔG is negative: the cell reaction proceeds spontaneously.

If the components of the cell are present at standard conditions,

$$\Delta G° = -nFE°$$

Voltage and Concentrations

The voltage of the cell:

$$^- \text{Zn(s)} \mid \text{Zn}^{2+}\text{(aq, 1.0 } M) \mid\mid \text{Cu}^{2+}\text{(aq, 1.0 } M) \mid \text{Cu(s)} ^+$$

with all the constituents in their standard states (and ignoring any contribution of the salt bridge to the cell voltage), is

$$E° = 1.10 \text{ volts}$$

The cell reaction is:

$$\text{Zn(s)} + \text{Cu}^{2+}\text{(aq)} \rightarrow \text{Zn}^{2+}\text{(aq)} + \text{Cu(s)}$$

If the cell voltage is a measure of the tendency for a reaction to occur, how will changing the concentrations of copper (II) ion and a zinc ion affect the cell voltage?

From Le Chatelier's principle, we would expect that increasing the concentration of copper (II) ion, or decreasing the concentration of zinc ion, would increase the tendency for the cell reaction to occur. How can we calculate the effect on the cell voltage?

In Chapter 16 we used:

$$\Delta G = \Delta G° + RT \ln Q$$

where Q is the "reaction quotient," to calculate the effect of changing concentrations on ΔG.

Now we have:

$$\Delta G = -nFE$$

where E is the cell voltage for a given set of concentrations;

and
$$\Delta G° = -nFE°$$

where E° is the cell voltage when all the components are present in their standard states.

Then:

$$\Delta G = \Delta G^\circ + RT \ln Q$$

becomes:

$$-nFE = -nFE^\circ + RT \ln Q$$

or:

$$E = E^\circ - \frac{RT}{nF} \ln Q, \text{ the Nernst equation.}$$

At 25°C this becomes:

$$E = E^\circ - \frac{0.0592}{n} \log Q$$

Now we can calculate the cell voltage for any concentrations of reactants and products.

Problem 17-4:

What is the cell voltage of the following cell:

$$^- \; Zn(s) \mid Zn^{2+}(aq, \; 0.010 \; M) \parallel Cu^{2+}(aq, \; 1.00 \; M) \mid Cu(s) \; ^+$$

Solution:

Write the cell reaction:

$$Zn(s) + Cu^{2+}(aq) \rightarrow Cu(s) + Zn^{2+}(aq)$$

Calculate E°:

$$Cu^{2+} + 2e^- \rightarrow Cu(s) \qquad +0.34 \text{ V}$$

$$Zn^{2+} + 2e^- \rightarrow Zn(s) \qquad -0.76 \text{ V}$$

$$[\text{or } Zn(s) \rightarrow Zn^{2+} + 2e^- \qquad +0.76 \text{ V}]$$

Then E° = 1.10 V

Write the expression for Q for the cell reaction:

$$Q = \frac{[Zn^{2+}]}{[Cu^{2+}]} \qquad \begin{array}{l} [Cu(s) \text{ and } Zn(s) \text{ don't appear} \\ \text{in the expression.}] \end{array}$$

Calculate E:

$$E = E^\circ - \frac{0.0592}{n} \log Q = 1.10 \text{ V} - \frac{0.0592}{n} \log \frac{[Zn^{2+}]}{[Cu^{2+}]}$$

n = 2: when the cell reaction takes place as written, 2 moles of electrons are transferred. Now use the concentrations in the cell:

$$E = 1.10 \text{ V} - \frac{0.0592}{2} \log \frac{(0.10)}{(1.0)}$$

$$= 1.10 \text{ V} - \frac{0.0592}{2} \log (10^{-2})$$

$$= 1.10 \text{ V} - \frac{0.0592}{2} (-2)$$

$$= 1.10 \text{ V} + .06 \text{ V} = 1.16 \text{ V}$$

Notice that lowering the zinc ion concentration (the product of the cell reaction), increases the tendency for the cell reaction to occur.

Problem 17-5:

What is the cell voltage of the following cell:

$$Cu(s) \mid Cu^{2+}(aq, \; 0.01 \; M) \parallel Cu^{2+}(aq, \; 1.0 \; M) \mid Cu(s)$$

Solution:

Write the cell reaction with oxidation at the left electrode:

$$Cu(s) + Cu^{2+}(aq, \; 1.0 \; M) \rightarrow Cu^{2+}(aq, \; 0.01 \; M) + Cu(s)$$

E° for this cell is 0.00V: the two reduction half-reactions are the same: $Cu^{2+} + 2e^- \rightarrow Cu$.

$$E = E° - \frac{0.0592}{2} \log \frac{[Cu^{2+}] \text{ product}}{[Cu^{2+}] \text{ reactant}} = E° - \frac{0.0592}{2} \log \frac{[Cu^{2+}] \text{ left}}{[Cu^{2+}] \text{ right}}$$

$$= 0.00 - \frac{0.0592}{2} \log \frac{[0.01]}{[1.0]}$$

$$= 0.00 - \frac{0.0592}{2} \log 10^{-2}$$

$$= 0.00 - \frac{0.0592}{2} (-2)$$

$$= +0.0592 \text{ V}$$

This overall cell reaction is spontaneous, as Le Chatelier's principle would predict. This cell is a "concentration cell."

Cell Voltages and Equilibrium Constants

For any cell:

$$E = E° - \frac{RT}{nF} \ln Q$$

At 25°C

$$E = E° - \frac{.0592}{n} \log Q$$

When our cell is at equilibrium: $E = 0$ and no further reaction takes place. Then:

$$0 = E° - \frac{RT}{nF} \ln K$$

$$\text{or} \quad E° = \frac{+RT}{nF} \ln K$$

$$\text{at 25°C:} \quad E° = \frac{.0592}{n} \log K$$

Problem 17-6:

What is the equilibrium constant for the reaction:

$$Zn(s) + Cu^{2+}(aq) \rightarrow Cu(s) + Zn^{2+}(aq)$$

Solution:

Write the overall reaction in terms of half-reactions: find E° for each from Table 17.1:

$$Zn \rightarrow Zn^{2+} + 2e^- \qquad E° = +0.76 \text{ V}$$

$$Cu^{2+} + 2e^- \rightarrow Cu \qquad E° = +0.34 \text{ V}$$

Calculate E° for the overall reaction:

$$E° = +0.76 \text{ V} + 0.34 \text{ V} = 1.10 \text{ V}$$

Calculate K:

$$E° = \frac{.0592}{n} \log K$$

$$1.10 \text{ V} = \frac{.0592}{n} \log K$$

$$\log K = \frac{2(1.10)}{.0592} = 37.16$$

$$K = 10^{37.16} = 1.4 \times 10^{37}$$

This reaction goes very nearly to completion!

Electrolysis

A galvanic cell harnesses chemical energy to perform electrical work. An electrolytic cell uses electrical energy to perform "chemical work," to carry out a non-spontaneous chemical process.

We have already used:

$$q = i \cdot t \quad \text{and} \quad n = \frac{q}{F}$$

to calculate how much copper would be plated out if a current were passed through a solution of copper (II) ion.

If we electrolyze a solution with a mixture of ions, we can use the reduction potentials to determine which species is most easily reduced, and the oxidation potentials to determine which species is most easily reduced.

Problem 17-7:

A voltage is applied to an aqueous solution of $NiCl_2$ and $CdCl_2$.

What reaction occurs at the cathode? What reaction occurs at the anode?

First, list the principal species in solution:

$$Ni^{2+}, \quad Cd^{2+}, \quad Cl^-, \quad H_2O$$

Reduction occurs at the cathode: Use Table 17.1 to find the possible reductions:

$$Ni^{2+} + 2e^- \rightarrow Ni \qquad -0.14 \text{ V}$$

$$Cd^{2+} + 2e^- \rightarrow Cd \qquad -0.40 \text{ V}$$

$$H_2O + 2e^- \rightarrow H_2 + 2 OH^- \qquad -0.83 \text{ V}$$

Nickel (II) ion is the most easily reduced and will plate out on the cathode first. When the nickel ion is depleted, cadmium (II) will begin to plate.

Oxidation occurs at the anode: Use Table 17.1 to find the possible oxidations:

$$2 H_2O \rightarrow O_2 + 4 H^+ + 4e^- \qquad -1.23 \text{ V}$$

$$2 Cl^- \rightarrow Cl_2 + 2e^- \qquad -1.36 \text{ V}$$

H_2O should be more easily oxidized than Cl^-; however, when we electrolyze this solution, Cl_2 is produced first. These electrode potentials are thermodynamic values; an excess voltage (overvoltage) is required to make some reactions proceed at a reasonable rate. The overvoltage required to produce O_2 is large enough that Cl^- is oxidized before H_2O is oxidized.

17-1.   Use the data in Table 17.1 to choose the stronger oxidizing agent in each pair.

    a. Cl_2 or Ag^+

    b. Zn^{2+} or Sn^{2+}

    c. Cu^{2+} or Br_2

17-2.   Use the data in Table 17.1 to choose the stronger <u>reducing</u> agent in each pair.

    a. Fe^{2+} or Cu

    b. Ag or Zn

17-3.   Consider a cell, with two compartments; one with a copper electrode dipping into 1.0 M $Cu(NO_3)_2$, and the other with a silver electrode dipping into 1.0 M $AgNO_3$.

    a.   Which electrode is the anode (where oxidation occurs)?

    b.   Which electrode is the positive electrode (towards which electrons flow in the external circuit)?

    c.   What is E° for the cell?

17-4.   A 2.50 amp current is passed through a solution of $Ni(NO_3)_2$ for 90.0 minutes. How many grams of nickel is deposited on the cathode?

17-5.   For the cell

    $- Zn(s) \mid Zn^{2+}(aq, 1.0\ M) \mid\mid Cu^{2+}(aq, 1.0\ M) \mid Cu +$ E° = 1.10 V

    a.   Write the cell reaction; with oxidation at the left electrode.

    b.   Using Le Chatelier's principle, will the cell voltage increase or decrease if we <u>reduce</u> the concentration of Zn^{2+} in the left hand compartment.

    c.   Will the cell voltage increase or decrease if we <u>reduce</u> the concentration of Cu^{2+} in the right hand compartment.

    d.   Calculate the cell voltage when $[Zn^{2+}]$ = 0.10 M and $[Cu^{2+}]$ = 0.50 M.

17-1.   The stronger <u>oxidizing</u> <u>agent</u> is the more easily <u>reduced</u>, and has the most positive <u>reduction</u> potential.

   a. Cl_2

   b. Sn^{2+}

   c. Br_2

17-2.   The stronger <u>reducing</u> <u>agent</u> is the more easily <u>oxidized</u> and has the most positive <u>oxidation</u> potential.

   a. $Fe^{2+} \rightarrow Fe^{3+} + e^-$ $E^{\circ}_{ox} = -0.77$ V

      $Cu \rightarrow Cu^{2+} + 2\ e^-$ $E_{ox} = -0.34$ V

      Cu is more easily oxidized.

   b. $Ag \rightarrow Ag^+ + e^-$ $E^{\circ}_{ox} = -0.80$ V

      $Zn \rightarrow Zn^{2+} + 2\ e^-$ $E^{\circ}_{ox} = +0.76$ V

      Zn is more easily oxidized.

17-3.   Write the half-reactions:

   $Cu^{2+} + 2\ e^- \rightarrow Cu(s)$ $E^{\circ} = +0.34$ V

   $Ag^+ + e^- \rightarrow Ag(s)$ $E^{\circ} = +0.80$ V

   a. Ag^+ is more easily reduced than Cu^{2+};

      Cu is more easily oxidized than Ag.

      Oxidation will occur at the copper electrode.

   b.   Electrons will flow from the copper electrode <u>to</u> the silver electrode.   The silver electrode is the positive elec<u>t</u>rode.

   c. $E^{\circ}_{cell} = E^{\circ}_{Ag^+/Ag} - E^{\circ}_{Cu^{2+}/Cu}$

      $= 0.80$ V $- (0.34$ V$) = 0.46$ V

   We can write this cell as:

      $-Cu \mid Cu^{2+}$ (1.0 M) $\mid\mid Ag^+$ (1.0 M) $\mid Ag +$

17-4. Charge $= i \cdot t$

$$= 2.50 \text{ C/s} \times 90.0 \text{ min} \times 60.0 \text{ s/min} = 1.35 \times 10^4 \text{ C}$$

Moles of electrons $= \dfrac{1.35 \times 10^4 \text{ C}}{96,485 \text{ C/mole e}^-} = 0.140$ moles

$Ni^{2+} + 2 \text{ e}^- \rightarrow Ni$:

Moles of Ni $= 1/2$ (moles of electrons)

$$= 0.0700 \text{ moles}$$

Mass of Ni $= (7.00 \times 10^{-2} \text{ moles}) \times 58.7 \text{ g/mole}$

$$= 4.11 \text{ g}$$

17-5. a. $Zn + Cu^{2+} \rightarrow Zn^{2+} + Cu$

b. The tendency for the cell reaction to occur should <u>increase</u>; E_{cell} increases.

c. The tendency for the cell reaction to occur should <u>decrease</u>; E_{cell} decreases.

d. $E_{cell} = E° - \dfrac{0.0592}{n} \log \dfrac{[Zn^{2+}]}{[Cu^{2+}]}$

$$= 1.10 \text{ V} - \dfrac{0.0592}{2} \log \dfrac{(0.10)}{(0.50)}$$

$$= 1.10 \text{ V} + 0.02 \text{ V} = 1.12 \text{ V}$$

CHAPTER EIGHTEEN: REPRESENTATIVE ELEMENTS I – GROUPS 1A TO 4A

Chapters 18, 19 and 20 focus on the "descriptive chemistry" of groups of elements. In the course of the first seventeen chapters, and your work in the laboratory, you have been exposed to a great deal of chemistry. These chapters collect much of the information by chemical "family" and offer you a chance to apply the chemical principles you've learned, principles of bonding, thermodynamics, molecular structure, intermolecular forces, and electro-chemistry to these families. If you're unsure of these principles, make a quick review when you meet the concepts again in these chapters.

Remember the general trends:

Atomic size: <u>increases</u> down a column in the periodic table, and <u>decreases</u> <u>across a row</u> in the periodic table moving from left to right.

First ionization energy: <u>decreases</u> down a column and <u>increases</u> across a row in the periodic table moving from left to right.

Electronegativity: <u>decreases</u> down a column and increases across a row in the periodic table, moving from left to right.

Now, if we look a bit more closely we find differences, as well as similarities in a family, and exceptions to the trends.

In each family, look for trends within the family, and elements which don't follow a trend. Study the summary tables of reactions for elements in a family; then, write equations for the reactions. Your challenge is to understand as well as memorize. Keep your periodic table at hand.

Some useful principles to review:

- Elements tend to lose valence electrons, or gain valence electrons for form a "complete" outer valence shell. Elements can complete valence shells by becoming ions, or sharing electrons.

- The elements in Row 2 can form π bonds; elements beyond Row 2 don't form effective π bonds. For example, nitrogen forms diatomic molecules: $N \equiv N$, while phosphorus forms tetrahedral P_4 molecules with no multiple bonds.

- Metallic character increases down a column in the periodic table.

- Electronegativity increases to the right in the periodic table.

Now: Look at the first ionization energies of some elements in Row 3.

I.E. (kJ/mol)

Na	495
Mg	735
Al	580
Si	780
P	1060
S	1005
Cl	1255
Ar	1525

As the nuclear charge increases across the row, the ionization energy increases, with two exceptions.

Why is the ionization energy for aluminum less than that for magnesium? Write the outer electron configurations for each atom:

Mg   [Ne] $3s^2$

Al   [Ne] $3s^2 3p^1$

Magnesium "loses" a 3s electron; aluminum "loses" a 3p electron. The 3p electron is shielded from the nucleus somewhat by the 3s electrons: 3p orbitals in a multi-electron atom are higher in energy than 3s orbitals. Therefore, even though the nuclear charge is higher in aluminum, the 3p electron in aluminum is more easily "lost" than the 3s electron in magnesium.

Why is the ionization energy for sulfur less than that for phosphorus? Again, write the outer electron configurations:

P   [Ne] $3s^2 3p^3$

S   [Ne] $3s^2 3p^4$

Each atom "loses" a 3p electron; however, the electron "lost" by phosphorus occupies a "half-filled" orbital. The electron "lost" by sulfur occupies a "filled" orbital. It is easier to remove an electron from a filled orbital because of the repulsion between two electrons sharing the same orbital.

Other differences: non-metals in Row 2 can form π bonds, while elements in Row 3 don't form effective π bonds. On the other hand, elements in Row 2 don't use "expanded octets" in bonding, while elements in Row 3 and beyond often use "expanded octets" in bonding. How do we explain these phenomena? The atoms in Row 3 are larger in size than the atoms in Row 2. The "sideways" overlap of p orbitals is less effective between two larger atoms. Therefore, elements in Row 3 don't form effective π bonds. Atoms in Row 3 do have d orbitals, as well as s and p orbitals, to use in bonding; hence, these atoms have the ability to use "expanded octets" in bonding.

275

These differences in bonding lead to differing chemistry for the atoms in the same family.

As you study the chemistry of the representative elements, "compare-and-contrast" the chemistry of elements in the same family; look for similarities and differences. Look for trends and deviations from the trends. Try to puzzle out the differences.

For example, we might expect a trend in the oxidation potential ($E°_{ox}$) for an "alkali metal" (Li, Na, K, Rb, Cs) to follow the ionization energy. Compare:

$$Na(s) \rightarrow Na^+(aq) + e^- \qquad E°_{ox}$$

and

$$Na(g) \rightarrow Na^+(g) + e^-(g) \qquad I.E.$$

Element	I.E. kJ/mol	Std. Oxidation Potential (V)	Std. Reduction Potential (V)
Li	520	+3.05 V	−3.05 V
Na	496	+2.71	−2.71
K	419	+2.92	−2.92
Rb	409	+2.99	−2.99
Cs	382	+3.02	−3.02

The oxidation potential for lithium doesn't fit the trend. Why? The half-reaction

$$Li(s) \rightarrow Li(aq) + e^-$$

is more complex than the removal of an electron from the gaseous atom. We can break the overall half-reaction into a series of simpler steps:

$$Li(s) \rightarrow Li(g) \qquad \text{The energy of sublimation (+)}$$

$$Li(g) \rightarrow Li^+(g) + e^-(g) \qquad \text{The ionization energy (+)}$$

$$Li^+(g) \xrightarrow{H_2O} Li^+(aq) \qquad \text{The hydration energy (−)}$$

The ionization energy of lithium is larger than the ionization energy of sodium; that is, it takes more energy to remove an electron from a gaseous lithium atom than from a gaseous sodium atom. However, the lithium ion is much smaller than the sodium ion. Therefore, the lithium ion interacts more strongly with water molecules when it is hydrated. The lithium ion's large negative hydration energy compensates for lithium's larger ionization energy, so that lithium is more easily oxidized than sodium

How can you organize "descriptive" chemistry of a "family" of elements? You might want to look at trends in "physical" properties (such as melting point, boiling point, or density) and "chemical" properties (such as reduction potentials).

Then <u>write</u> equations for each element for:

- The methods of preparation; are there similarities for different elements in the same family?

- Reactions with oxygen; are the oxides acidic or basic? Are the oxides ionic or covalent? Does the element react to form peroxides? Superoxides?

- Reactions with hydrogen; are these compounds covalent or ionic?

- Reactions with water, with halogens, with nitrogen, with acid solutions.

Can you explain any differences or trends? Practice naming the compounds formed. Compare and contrast the non-metallic or metallic character of each element with others in the same family. For metals, compare the solubility of their salts (carbonates, chlorides, sulfates and sulfides).

You'll find that the problems in the text draw on many of the concepts you've previously learned.

CHAPTER EIGHTEEN: SELF-TEST

18-1.   Complete and balance the following reactions:

a.   ___ Na(s) + ___ H$_2$O(ℓ) → ___ H$_2$(g) + _____(aq)

b.   ___ Mg(s) + ___ O$_2$(g) → _____(s)

18-2.   Write the formula for each of the following compounds:

a.   Aluminum oxide

b.   Sodium nitrate

c.   Magnesium sulfate

d.   Potassium sulfide

18-3.   CO$_2$ is a gas; SiO$_2$ is a brittle solid.   Explain the differences in these oxides.

18-4.   In each pair, choose the species with the larger ionization potential. (From which is it more difficult to remove an electron?)

a.   Na or Mg

b.   Na or K

c.   Mg or K

d.   Na or Na$^+$

18-5.   In each set, choose the species with the larger radius.

a.   Na or Rb

b.   Mg or K

c.   Mg^{2+} or Na$^+$

18-6.   Are oxides of IA and IIA elements acidic or basic?

18-7.   Are the compounds of IA and IIA elements with hydrogen covalent or ionic?

CHAPTER EIGHTEEN: SELF-TEST ANSWERS

18-1. a. $2\ Na(s) + 2\ H_2O(\ell) \rightarrow 2\ NaOH(aq) + H_2(g)$

b. $2\ Mg(s) + O_2(g) \rightarrow 2\ MgO(s)$

18-2. a. Al_2O_3

b. $NaNO_3$

c. $MgSO_4$

d. K_2S

18-3. CO_2 molecules are non-polar. SiO_2 is a network solid; there are no molecules as such in $SiO_2(s)$.

18-4. a. Mg > Na

b. Na > K

c. Mg > Na > K

d. Na^+ > Na

18-5. a. Rb > Na

b. K > Na > Mg

c. $Na^+ > Mg^{2+}$; Na^+ and Mg^{2+} are isoelectronic.

18-6. Basic: $Na_2O(s) + H_2O(\ell) \rightarrow 2\ Na(OH)(aq)$

18-7. Ionic, such as NaH, and CaH_2

Again, we can use the approach we developed in Chapter 18: look for trends in physical properties and chemical properties for each family of elements.

Remember, metallic character increases as we go down a column in the periodic table. This is most marked in Columns 5A and 6A. For example, nitrogen and phosphorus are non-metallic, forming acidic oxides, and covalent compounds with hydrogen. Bismuth and antimony do lose electrons to form cations in some compounds. Bismuth and antimony also form covalent compounds, such as, BiF_5 and $SbCl_5$.

Notice the differences between the chemistry of nitrogen, which has the ability to form strong π bonds, and the chemistry of phosphorus which doesn't form π bonds but can have an "expanded octet," as in PF_5.

Remember that when a hydrogen atom is covalently bound to a small electronegative atom (nitrogen, oxygen, or fluorine), a "hydrogen bond" can be formed with another small electronegative atom with a "lone pair" of electrons.

Problem 19-1:

Why does nitrogen occur as N_2, while phosphorus occurs as P_4? Why does oxygen occur as O_2, while sulfur occurs as S_8?

Solution:

Draw the Lewis structure for each molecule:

N_2:   :N $\equiv$ N: P_4:

O_2 :O $=$ O: S_8:

Phosphorus cannot form the strong π bonds that would be required in P_2; even though the P-P bonds are "strained" in P_4, P_4 contains only single bonds. Similarly, sulfur is too large to form the π bonds that would be required in S_2; sulfur forms rings and chains.

Problem 19-2:

Why is the boiling point of AsH_3 higher than that for PH_3? Why is the boiling point of NH_3 higher than that for either PH_3 or AsH_3?

Solution:

To understand the boiling points, we need to determine the intermolecular forces, the forces between these molecules. First, write the Lewis structure for each molecule:

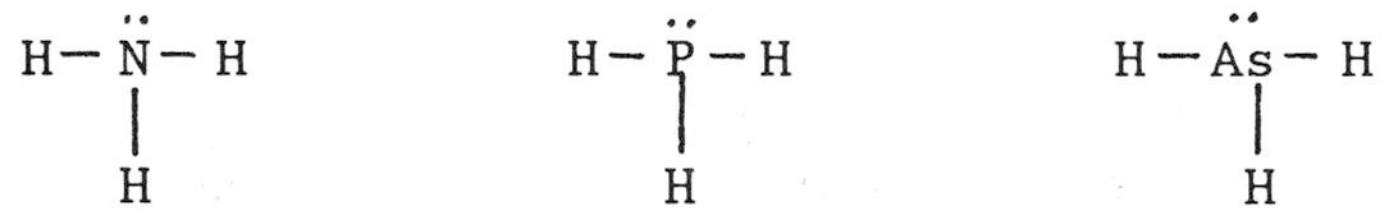

Are the bonds in these molecules polar? Look at the electronegativities:

	E.N.
H	2.1
N	3.0
P	2.1
As	2.0

Only the N − H bonds are polar: PH_3 and AsH_3 are non-polar molecules. Is NH_3 polar? Look at the molecular geometry: use VSEPR:

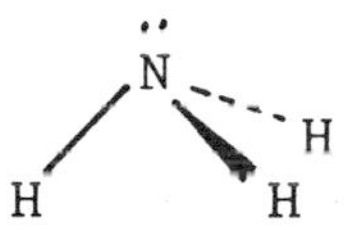

NH_3 is a pyramidal polar molecule. Notice, NH_3 can form hydrogen bonds: each hydrogen atom is bound to a small, electronegative nitrogen and each nitrogen has a lone pair of electrons.

These hydrogen bonds are much stronger than the London dispersion forces between non-polar molecules. Therefore, NH_3 has the highest boiling point of these three compounds.

Remember that London dispersion forces, the attractions due to "transient dipoles," increase with an increase in the number of electrons in the molecule. Therefore, AsH_3 has a higher boiling point than PH_3.

Problem 19-3:

Why is ozone (O_3) "v-shaped" (bent) and not linear?

Solution:

Draw the Lewis structure and use VSEPR to predict the geometry:

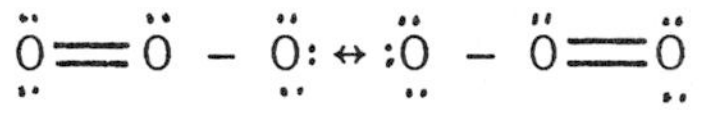

There are three "sets" of electrons around the central oxygen atom: Therefore, the O − O − O bond angle will be about 120°:

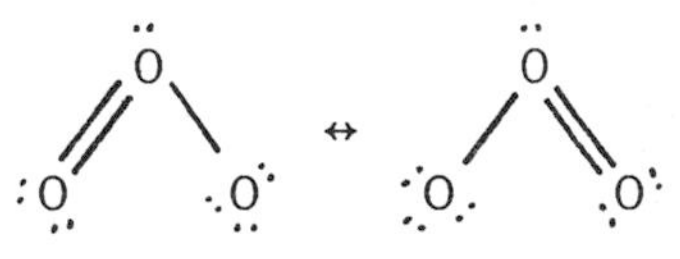

Problem 19-4:

The oxygen-oxygen single bond length (as in hydrogen peroxide) is approximately 0.149 nm. The oxygen-oxygen double bond length (as in O_2) is 0.121 nm. How long would you expect the oxygen-oxygen bonds in ozone to be?

Solution:

From the resonance structures in 19-3, we would expect each of the oxygen-oxygen bonds in ozone to be "one-and-a-half" bonds. We'd expect the O - O bond length to be approximately midway between the single bond length and the double bond length, or about 0.135 nm.

The Halogens

Notice the increase in melting point and boiling point as the halogen (X_2) molecules become heavier. Look at the trends in oxidizing power of the halogens, with F_2 the most easily reduced, and the strongest oxidizing agent.

The discussion of the acid strength of HF reminds us of the importance of both enthalpy and entropy changes in determining whether a process is favored. Although $\Delta H°$ for the dissociation of HF is favorable, $\Delta S°$ for the dissociation is unfavorable. $\Delta G°$ for the dissociation of HF is positive, and K_a is small. HF is a weak acid.

Look at the many oxyacids of the halogens. Practice naming the acids and their corresponding salts. Remember, the prefix hypo- refers to an acid with fewer oxygens than the acid ending in -ous; the prefix per- refers to an acid with more oxygens than the acid ending in -ic.

Therefore, we have:

hypochlorous acid	HOCl		
chlorous acid	HOClO	or	$HClO_2$
chloric acid	$HOClO_2$	or	$HClO_3$
perchloric acid	$HOClO_3$	or	$HClO_4$

Chlorine, can disproportionate. In a disproportionation reaction, an element is both oxidized and reduced.

$$Cl_2(aq) + H_2O(l) \rightleftharpoons HOCl(aq) + H^+(aq) + Cl^-(aq)$$

Write the two half-reactions:

$$2\ H_2O + Cl_2 \rightarrow 2\ HOCl + 2\ H^+ + 2e^-: Cl_2 \text{ is oxidized}$$

$$2e^- + Cl_2 \rightarrow 2\ Cl^-: Cl_2 \text{ is reduced}$$

Remember, write (and balance) as many reactions as possible.

CHAPTER NINETEEN: SELF-TEST

19-1. Are the oxides of 5A, 6A and 7A acidic or basic? Give an example for each family.

19-2. Are the compounds of 5A, 6A and 7A elements with hydrogen covalent or ionic? Give an example from each family.

19-3. Name each of the following compounds:

a. HNO_2

b. H_2SO_4

c. HBr

d. NO_2

19-4   a. Why is the boiling point of HBr higher than that of HCl?

b. Why is the boiling point of HF higher than that of HCl?

19-5. In each pair, choose the species with the larger ionization potential.

a. O or O^+

b. O or F

c. F or Cl

19-6. In each pair, choose the species with the larger radius.

a. N^{3-} or O^{2-}

b. F^- or Ne

c. O or O^{2-}

d. O or S

19-7. In each pair, choose the atom with the greatest tendency to gain an electron.

a. O or F

b. S or O

c. Se or Cl

19-8. Write balanced equations for:

a. The reaction of fluorine with sodium to form sodium fluoride

b. Burning graphite to yield carbon monoxide

283

19-9. Balance the following equations in acidic solution.

    a. ___ Cu + ___ NO_3^- → ___ Cu^{2+} + ___ NO

    b. ___ H_2O_2 + ___ I^- → ___ I_2 + ___ H_2O

    c. ___ I_2 + ___ $S_2O_3^{2-}$ → ___ $S_4O_6^{2-}$ + ___ I^-

    d. ___ Cl_2 → ___ Cl^- + ___ ClO_3^-

CHAPTER NINETEEN: SELF-TEST ANSWERS

19-1. Acidic; $N_2O_5 + H_2O \rightarrow 2\ HNO_3$

$$H_2O + SO_3 \rightarrow H_2SO_4$$

$$Cl_2O + H_2O \rightarrow 2\ HClO$$

19-2. Covalent; NH_3, H_2O, HF

19-3. a. <u>nitrous</u> acid

b. <u>sulfuric</u> acid

c. hydrobromic acid

d. nitrogen dioxide

19-4. a. Stronger London dispersion forces in HBr

b. Hydrogen bonding in HF

19-5. a. O^+

b. F

c. F

19-6. a. N^{3-}; O^{2-} is isoelectronic with a larger nuclear charge.

b. F^-

c. O^{2-}

d. S

19-7. a. $F > O$

b. $O > S$

c. $Cl > S > Se$

19-8. a. $F_2(g) + 2\ Na(s) \rightarrow 2\ NaF(s)$

b. $2\ C(gr) + O_2(g) \rightarrow 2\ CO(g)$

19-9. a. $3\ Cu + 8\ H^+ + 2\ NO_3^- \rightarrow 2\ NO + 4\ H_2O + 3\ Cu^{2+}$

b. $2\ H^+ + 2\ I^- + H_2O_2 \rightarrow 2\ H_2O + I_2$

c. $I_2 + 2\ S_2O_3^{2-} \rightarrow S_4O_6^{2-} + 2\ I^-$

d. $3\ H_2O + 3\ Cl_2 \rightarrow 5\ Cl^- + ClO_3^- + 6\ H^+$

CHAPTER TWENTY: TRANSITION METALS AND COORDINATION CHEMISTRY

The transition elements (or transition metals) are the elements found in the "d-block" and the "f-block" of the periodic table. As we "build" the electron configurations for these atoms, the last electrons we add are d or f electrons.

Trends

We still find similarities in chemistry within a group (column), but we find more similarities between groups (columns) of transition metals than we do between groups of representative elements.

Size

Atomic size of transition metals decreases to the right in a period (row) from group III B through VIII B, as the nuclear charge increases. Atoms in the 4d row are larger than their counterparts in the 3d row. However, atoms in the 5d row are nearly the same size as those in the 4d row. The lanthanides, the 4f row, occur between the 4d row and the 5d row. There is a steady decrease in size through the lanthanide row, again, with increasing nuclear charge. This "lanthanide contraction" offsets the increase in size we would expect between the 4d and 5d rows.

Oxidation Potentials

The "oxidation potential" is a measure of the tendency for a metal to be oxidized, the tendency for a metal to act as a reducing <u>agent</u>. The standard oxidation potential for a metal, M, is E° for the reaction:

$$M(s) + 2\ H^+(aq) \rightarrow H_2(g) + M^{2+}(aq)$$

$$\text{or } 2\ M(s) + 6\ H^+(aq) \rightarrow 3\ H_2(g) + 2\ M^{3+}(aq)$$

Table 20.3 shows that transition metals generally become more difficult to oxidize as we go from left to right in the periodic table. Chromium and zinc don't follow the trend.

Problem 20-1:

Make a plot of "oxidation potential" versus atomic number for the 3d transition elements. On the y-axis plot oxidation potential in volts; on the x-axis plot the elements: Sc, Ti, V, Cr....

Often a graph can give us a powerful picture of a trend, and its exceptions.

Oxidation States

Transition metals commonly form +2 and +3 ions. In general, transition metals form many oxidation states (see Table 20.2), although the higher oxidation states cannot exist as hydrated ions. The highly charged metal ion causes the water molecules bound to the metal to be acidic, losing protons. Instead of $V(H_2O)_n^{+5}$ we find VO_2^+; instead of $V(H_2O)_n^{+4}$ we find VO^{2+}.

Electron Configurations

We have used the periodic table as a guide to writing electron configurations. For the "d-block" transition metals, d orbitals are being filled. Scandium has the electron configuration $4s^2 3d^1$, indicating that the 4s orbital is lower in energy than the 3d orbital. Chromium and copper have slightly irregular electron configurations, $4s^1 3d^5$ and $4s^1 3d^{10}$. We find more irregularities in the lanthanides, the actinides and the 5-d transition metal row, as the orbitals are very close in energy.

In forming ions, transition metals appear to lose the valence shell electrons first. (We don't know which electrons are actually lost. We do know that in the ground state the ions have no outer s electrons.) Then d electrons are lost, to form the ion of the desired charge.

Problem 20-2:

Write the electron configuration for each of the following ions:

a. Cu^{2+}

b. Zn^{2+}

c. Fe^{3+}

d. Cr^{3+}

e. V^{2+}

Solution:

i) Write the electron configuration for the atom.

ii) Remove the outer s electrons first.

iii) Remove d electrons to form the desired ion.

a. Cu: $[Ar]3d^{10}4s^1$

removing two electrons: Cu^{2+}: $[Ar]3d^9$

b. Zn: $[Ar]3d^{10}4s^2$ Zn^{2+}: $[Ar]3d^{10}$

c. Fe: $[Ar]3d^6 4s^2$ Fe^{3+}: $[Ar]3d^5$

d. Cr: $[Ar]3d^5 4s^1$ Cr^{3+}: $[Ar]3d^3$

e. V: $[Ar]3d^3 4s^2$ V^{2+}: $[Ar]3d^3$

Naming Ionic Compounds

When we name a compound containing a metal ion which can have only one oxidation state, we don't need to indicate the oxidation state. When we name a compound containing a metal ion which can have more than one oxidation state, we indicate the oxidation state with a Roman numeral, in parentheses after the name of the metal.

For example:

$ZnCl_2$ Zinc chloride: Zinc forms only +2 ions

$FeSO_4$ Iron (II) sulfate: Iron forms +2 and +3 ions

$Fe(OH)_3$ Iron (III) hydroxide

Ag_2SO_4 Silver sulfate: Silver forms only +1 ions.

Colors

Ions that have incomplete "d shells" are often colored. The color of the ion depends both on the metal and the groups surrounding metal ion.

For example:

$Co(H_2O)_6^{2+}$ pink

$Cu(H_2O)_6^{2+}$ blue

$Ni(H_2O)_6^{2+}$ green

Zinc ion, Zn^{2+}, and silver ion, Ag^+, are both colorless, because they each have filled d shells.

Coordination Compounds

Coordination compounds typically contain a complex ion, a transition metal ion with other groups, ligands, bound to it. Examples of complex ions include:

$Co(NH_3)_6^{3+}$

$Cu(NH_3)_4^{2+}$

$Cr(H_2O)_6^{3+}$

$$PtCl_4{}^{2-}$$

$$Ni(CN)_4{}^{2-}$$

The coordination number for a metal ion is the number of ligands bound to the metal ion. Table 20.12 gives typical coordination numbers for some transition metal ions. Ions of higher charge tend to have higher coordination numbers.

Some complex ions are "inert" (inert to substitution), that is, the ligands are held tightly and exchange slowly with other molecules in the solution. Other complex ions are "labile," exchanging their ligands rapidly with the solution. The cobalt (III) complexes studied by Alfred Werner are inert. For example, if a solution of silver nitrate is added to a solution of $[Co(NH_3)_5Cl]Cl_2$ the following reaction occurs:

$$2\ Ag^+ + [Co(NH_3)_5Cl]^{+2} + 2\ Cl^- \rightarrow 2\ AgCl \downarrow + [Co(NH_3)_5Cl]^{2+}$$

The chloride ion bound directly to the cobalt (III) is not precipitated by the silver ion. The two chloride "counter-ions" were not bound to the cobalt (III) and are precipitated by the added silver ion.

Ligands

A ligand is a molecule or ion with a lone pair of electrons that can bond to a metal atom. If a molecule has two separate lone pairs of electrons, it may be able to occupy two positions in the coordination complex. Examples of bi-dentate ligands are oxalate, and ethylenediamine, as shown in Table 20.13. EDTA (ethylenediaminetetraacetate) can occupy six positions in a coordination complex.

Naming Complex Compounds

1. Name the cation before the anion.

2. In a complex ion, name the ligands before the metal ion.

3. In naming ligands:

   i) Add – o to the root name of anions:

       Chloro-, bromo-, hydroxo-, etc.

   ii) Use aquo for H_2O; ammine for NH_3; carbonyl for CO; nitrosyl for NO;

   iii) Use the name of the ligand for other neutral ligands.

   iv) Name the ligands alphabetically.

4. Use prefixes to denote the number of ligands of the same kind present:

    i) mono-, di-, tri-, tetra-, penta-, and hexa- for simple ligands.

    ii) bis-, tris-, tetrakis- for complex ligands.

5. Designate the oxidation state of the metal ion with a Roman numeral in parentheses.

6. If the complex ion is an anion, add the suffix -ate to the name of the metal.

Problem 20-3:

Name the following compounds:

a. $[Ni(NH_3)_4](ClO_4)_2$

b. $[Co(NH_3)_5Br]Cl_2$

c. $K_2[PtCl_6]$

Solution:

  i) Find the oxidation state of metal ion.

 ii) Then follow the naming rules.

a. NH_3 is a neutral ligand; perchlorate ion, ClO_4^- has a −1 charge. Therefore, nickel in this compound has a +2 oxidation state.

Name the cation first; then the anion

Tetraamminenickel(II) perchlorate

<u>Not</u> Tetraamminenickel(II) <u>di</u>perchlorate.

b. NH_3 is a neutral ligand; bromide and chloride each have a −1 charge. Therefore, cobalt in this compound has a +3 oxidation state.

Pentaamminebromocobalt(III) Chloride

c. Potassium ion has a +1 charge; chloride ion has a −1 charge. Therefore, platinum in this compound has a +4 oxidation state. The anion is $[PtCl_6]^{-2}$.

Potassium hexachloroplatin<u>ate</u> (IV)

Isomers

Isomers are compounds with the same formula, but different properties. The word "isomer" indicates a relationship between two compounds. We'll see many examples of isomers in the chapter on organic chemistry.

$[Co(NH_3)_5Br]SO_4$ and $[Co(NH_3)_5SO_4]Br$ are "coordination isomers." In the first compound bromide is bound directly to the cobalt ion; in the second compound bromide is a "counter ion," not bound directly to the cobalt ion. The compounds are examples of structural isomers.

These two molecules are structural isomers:

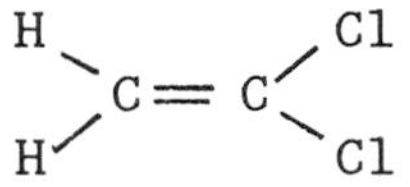

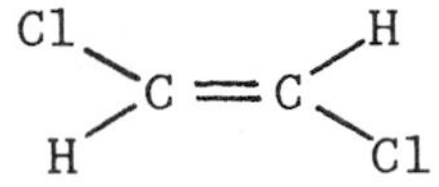

Stereoisomers

Stereoisomers have the same bonds in each molecule or ion, but a different spatial arrangement.

Geometrical isomers, or cis-trans isomers include the cis and trans pairs of $Pt(NH_3)_2Cl_2$ in Figure 20.11 and $Co(NH_3)_4Cl_2^+$ in Figure 20.12.

These two molecules are cis-trans isomers:

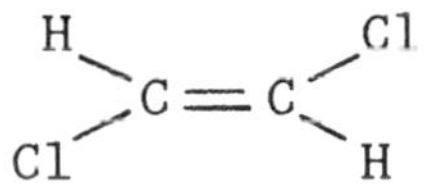

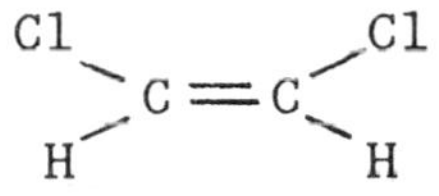

Optical isomer pairs are mirror images of each other, mirror images which cannot be superimposed. We can think of optical isomers as being "left-handed" and "right-handed." Tris (ethylenediamine) cobalt (III) ion can be either left- or right-handed.

The simplest example of a "chiral" molecule is a tetrahedral carbon atom with four different groups bound to it.

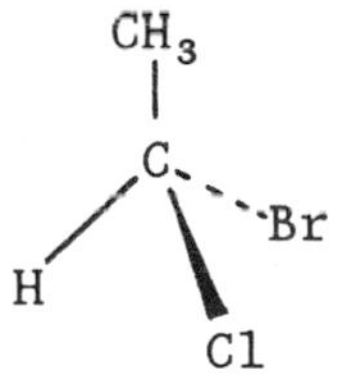

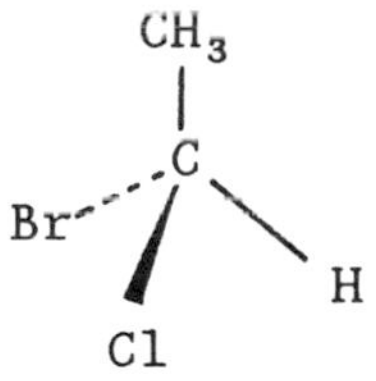

Build these two molecules with a model set. Convince yourself they are not superimposable.

You can determine whether a molecule is "optically active" by seeing if it is superimposable on its mirror image. If you find a "plane of symmetry" within a molecule, a plane that divides a molecule into two equivalent halves, each the mirror image of the other, the molecule is not optically active. Find the plane of symmetry in trans-$[Coen_2Cl_2]^+$ in Figure 20.18.

Bonding in Complex Ions

In the localized electron model of bonding in these ions, covalent bonds form when a filled ligand orbital, containing a lone pair of electrons, overlaps an empty hybrid orbital on the metal ion. The hybrid orbitals on the metal ion are formed from s, p and d orbitals:

d^2sp^3 hybrids to form octahedral complexes

dsp^2 hybrids to form square planar complexes

sp^3 hybrids to form tetrahedral complexes

sp hybrids to form linear complexes.

(The crystal field treats the ligands as negative charges about the metal ion.) This model focuses on the effect of the electric field created by the ligands on the energies of the d orbitals of the metal ion. In the isolated metal ion, the energies of all the d orbitals in a subshell are the same. The field due to the ligands "splits" the d orbitals into sets of differing energies.

Octahedral Complexes

Place the ligands along the x, y, and z axes. (Electrons which occupy d orbitals along the axes will be closer to the electrons in the ligand orbitals.) Therefore, d-orbitals which lie along the axes will be higher in energy than those which lie between the axes.

Which d orbitals lie along the axes? $d_{x^2-y^2}$ lies along the x and y axes; d_{z^2} lies along the z axis, with a "donut" in the xy plane. The d_{xy}, d_{xz}, and d_{yz} orbitals lie between the axes. The octahedral field "splits" and d orbitals into two sets: the d_{xy}, d_{yz}, and d_{xz} orbitals in one set and the $d_{x^2-y^2}$ and d_{z^2} in the other set. The sets are given "symmetry" labels: t_{2g} for the d_{xy}, d_{yz}, d_{xz} set and e_g for the $d_{x^2-y^2}$, d_{z^2} set.

The difference in energy between the two sets, the "crystal field splitting" is designated by Δ. The size of Δ depends both on the metal ion and the nature of the ligand.

Electron Configurations for Complex Ions

We can use the same rules that we've used before for determining electron configurations of complex ions:

i) Fill the available d orbitals with d electrons from the metal ion.

ii) Fill the lowest available orbitals first.

iii) Apply Hund's Rule: electrons will not "pair up" if there is another empty orbital of the same energy available.

iv)   The "crystal field splitting," Δ, can be smaller or larger than the "pairing energy," the energy of repulsion between two electrons in the same orbital.

iv$_a$)   If Δ is smaller than the pairing energy, electrons will half-fill the higher energy set of d orbitals before pairing.

iv$_b$)   If Δ is larger than the pairing energy, electrons will "pair up" in the lower energy set of d orbitals before filling the higher energy set.

Problem 20-4:

Give the d-electron configuration for:

a. $Ti(H_2O)_6^{3+}$ (Do you need to know the relative size of Δ?)

b. $V(H_2O)_6^{2+}$

c. $Fe(H_2O)_6^{2+}$ Δ is smaller than the pairing energy.

d. $[Fe(CN)_6]^{4-}$ Δ is larger than the pairing energy.

Solution

In each case, complexes are octahedral and the splitting diagram is:

$$\uparrow \quad \underline{} \quad \underline{} \quad e_g$$
$$\Delta$$
$$\downarrow \quad \underline{} \quad \underline{} \quad \underline{} \quad t_{2_g}$$

a. $Ti(H_2O)_6^{3+}$

Ti has the electron configuration: $[Ar]4s^2 3d^1$

Ti^{3+} has the electron configuration: $[Ar]3d^1$

We can write the d-electron configuration for $Ti(H_2O)_6^{3+}$ as: $(t_{2_g})^1$, that is, one electron in a t_{2_g} d orbital.

b. $V(H_2O)_6^{2+}$

V has the electron configuration: $[Ar]4s^2 3d^3$

V^{2+} has the electron configuration: $[Ar]3d^3$

There are 3 electrons to be placed in the d orbitals. We can place each of them in t_2 orbitals without pairing any electrons up.

Then the d-electron configuration for $V(H_2O)_6^{2+}$ is: $(t_{2_g})^3$

c. Fe(H$_2$O)$_6$]$^{2+}$

Fe has the electron configuration: [Ar]4s^{2}3d^6

Fe^{2+} has the electron configuration: [Ar]3d^6

There are 6 electrons to be placed in the d orbitals. We can place 3 in the t$_{2g}$ orbitals. The next electron must either pair up with an electron in a t$_{2g}$ orbital, or be placed in a higher energy e$_g$ orbital. In this case, Δ is smaller than the pairing energy. The lowest energy option available is to place the fourth electron in the e$_g$ orbital. The fifth electron will go in the remaining empty e$_g$ orbital. The sixth electron must go in a half-filled orbital; we'll put the sixth electron in a t$_{2g}$ orbital, the lowest energy half-filled orbital available. Our crystal field splitting diagram is:

$$\uparrow \qquad \uparrow \qquad e_g$$

$\uparrow$
Δ
$\downarrow$ $\quad \uparrow\downarrow \quad \uparrow \quad \uparrow \quad t_{2g}$

and the d-electron configuration is (t$_{2g}$)^4(e$_g$)^2

d. [Fe(CN)$_6$]$^{4-}$

Fe has the electron configuration [Ar]4s^{2}3d^6

Each cyanide ion has a –1 charge. Then the iron must have a +2 oxidation state.

Fe^{2+} has the electron configuration [Ar]3d^6

Again, there are six electrons to be placed in the d orbitals, the first three occupy the t$_{2g}$ orbitals. The next electron must either pair up with an electron in the t$_{2g}$ orbitals, or be placed in a higher energy e$_g$ orbital. In this case, Δ is larger than the pairing energy. The lowest energy option is to place the fourth, fifth and sixth electrons in the t$_{2g}$ orbitals.

Our crystal field splitting diagram is:

$$\underline{\quad} \qquad \underline{\quad} \qquad e_g$$

$\uparrow$
Δ
$\downarrow$ $\quad \uparrow\downarrow \quad \uparrow\downarrow \quad \uparrow\downarrow \quad t_{2g}$

In [Fe(H$_2$O)$_6$]$^{2+}$ there are 4 unpaired electrons; [Fe(H$_2$O)$_6$]$^{2+}$ is paramagnetic. In [Fe(CN)$_6$]$^{4-}$ there are no unpaired electrons; [Fe(CH)$_6$]$^{4-}$ is diamagnetic. When Δ is smaller than the pairing energy we have the "high spin" case, with the maximum number of unpaired electrons. When Δ is larger than the pairing energy we have the "low spin" case, with a smaller number of unpaired electrons.

Color and Complex Ions

Transition metal ions with unfilled d shells can absorb light, transferring an electron from a lower energy d orbital to a higher energy d orbital. The color of the solution is due to the light which is not absorbed. The simplest case to examine is a d^1 ion such as $Ti(H_2O)_6^{3+}$. The wavelength of light absorbed corresponds to the energy difference Δ. $Ti(H_2O)_6^{3+}$ absorbs light in the visible region with its absorption maximum at 510 nm. The solution removes yellow-green light from white light, leaving red and blue light. The solution appears violet.

Problem 20-5:

TiF_6^{3-} is an octahedral ion. F^- is a "weak field" ligand. Will the wavelength of light absorbed by TiF_6^{3-} be longer or shorter than the 510 nm wavelength absorbed by $Ti(H_2O)_6^{3+}$? What color do you expect the TiF_6^{3-} ion to be?

Solution:

The crystal field splitting, Δ, for TiF_6^{3-} will be smaller than that for $Ti(H_2O)_6^{3+}$. If the energy difference is smaller, the energy of the absorbed photon is smaller.

$$\Delta E = \frac{hc}{\lambda}$$

Therefore, TiF_6^{3-} will absorb light of a longer wavelength than 510 nm. Look at Table 20.16 in the text. If TiF_6^{3-} absorbs light of longer wavelengths, the ion will be blue to green.

The Molecular Orbital Model

The crystal field model describes interaction between ligands and the metal ion in a complex ion in terms of electrostatic repulsion. The molecular orbital model combines atomic orbitals on the metal atom with combinations of ligand orbitals to form bonding and anti-bonding molecular orbitals. The ligand orbitals containing the lone pairs are of σ symmetry.

The ligand orbitals lying along the x, y, and z axes can form bonding and anti-bonding orbitals with the s orbital, the p orbitals, and the d_{z^2} and $d_{x^2-y^2}$ orbitals on the metal atom. The d_{xy}, d_{yz}, and d_{xz} orbitals have the wrong symmetry to overlap the ligand σ orbitals: these d orbitals are non-bonding.

Look at the molecular orbital diagram for an octahedral complex in the text. The 6 lone pairs of electrons from the ligand orbitals will occupy the 6 bonding molecular orbitals: σ_s, σ_p and σ_d. The d electrons from the metal ion, then, will be placed in the non-bonding (t_{2g}) orbitals and the lowest anti-bonding (e_g^*) orbitals, depending on Δ, the difference in energy between these two sets of orbitals. The molecular orbital model predicts the same "splitting" pattern as the crystal field model. However, the splitting arises from a more realistic view of bonding in the complex.

295

CHAPTER TWENTY: SELF-TEST

20-1.   Write the electron configuration for each of the following ions:

    a. Ni^{+2}

    b. Ag^{+1}

    c. Zn^{+2}

    d. Sc^{+3}

20-2.   Which of the ions in 20-1 could be colored?

20-3.   Name the following compounds:

    a. Ag_2CO_3

    b. Fe_2O_3

    c. $SnCl_4$

    d. $CuCl$

20-4.   Name the following compounds:

    a. $K_3[Fe(CN)_6]$

    b. $[Cu(NH_3)_4]SO_4$

    c. $Na_2[MnCl_6]$

20-5.   Using the localized electron model for bonding in these complex ions, which hybrid orbitals does the metal ion use?

    a. $Ag(NH_3)_2{}^{+}$

    b. $Co(NH_3)_6{}^{3+}$

20-6.   Use the crystal field splitting diagram to determine the electron configuration for the following metal ions:

    a. $V(H_2O)_6{}^{2+}$

    b. $Co(CN)_6{}^{3-}$ Δ is larger than the pairing energy.

    c. $Fe(H_2O)_6{}^{3+}$ Δ is smaller than the pairing energy.

20-7. $[Pt(NH_3)_2Br_2]$ is a square planar complex. Draw the two isomers of this compound.

20-8. $[Co(CN)_6]^{3-}$ is a yellow ion, absorbing violet light. $[Co(NH_3)_6]^{3+}$ is an orange ion, absorbing blue light.

    a. Which complex absorbs light of longer wavelength?

    b. Which complex absorbs higher energy photons?

    c. Explain the difference in color of the ions in terms of crystal field theory.

20-9. $[Co(NH_3)_6]^{3+}$ is a diamagnetic ion; $[CoF_6]^{3-}$ is a paramagnetic ion. Give the electron configuration for the d electrons in each ion.

CHAPTER TWENTY: SELF-TEST ANSWERS

20-1. a. Ni^{2+}: $1s^2 2s^2 2p^6 3s^2 3p^6 3d^8 = [Ar]3d^8$

b. Ag^+: $[Kr]4d^{10}$

c. Zn^{2+}: $[Ar]3d^{10}$

d. Sc^{3+}: $[Ar]$

20-2. To be colored, an ion must have a partially filled d shell; of these ions, only Ni^{+2} is colored.

20-3. a. silver carbonate

b. iron(III) oxide

c. tin(IV) chloride

d. copper(I) chloride

20-4. a. potassium hexacyanoferrate(III)

b. tetramminecopper(II) sulfate

c. sodium hexachloromanganate(IV)

20-5. a. sp hybrid orbitals

b. d^2sp^3 hybrid orbitals

20-6. a.
$$\underline{\quad}\ \underline{\quad}\quad e_g$$
$$\underline{\uparrow}\ \underline{\uparrow}\ \underline{\uparrow}\quad t_{2g}$$

b.
$$\underline{\quad}\ \underline{\quad}\quad e_q$$
$$\underline{\uparrow\downarrow}\ \underline{\uparrow\downarrow}\ \underline{\uparrow\downarrow}\quad t_{2g}$$

c.
$$\underline{\uparrow}\ \underline{\uparrow}\quad e_g$$
$$\underline{\uparrow}\ \underline{\uparrow}\ \underline{\uparrow}\quad t_{2g}$$

20-7.
```
        NH₃                        NH₃

   Br - Pt - Br          NH₃ - Pt - Br

        NH₃                        Br

      trans                      cis
```

298

20-8. a. Blue light has a longer wavelength than violet light.

b. Violet light has higher energy photons than blue light.

c. CN^- causes a larger "crystal field splitting" than NH_3.

20-9.

___ ___ e_g	$\uparrow$ $\uparrow$ e_g
$\uparrow\downarrow$ $\uparrow\downarrow$ $\uparrow\downarrow$ t_{2g}	$\uparrow\downarrow$ $\uparrow$ $\uparrow$ t_{2g}
$[Co(NH_3)_6]^{3+}$	$[CoF_6]^{3-}$

CHAPTER TWENTY-ONE: THE NUCLEUS

In chemical reactions, the nuclei of the atoms remain unchanged, while electrons are rearranged, as bonds are broken and made, as ions form. Now we will look at nuclear reactions, reactions in which the nuclei themselves change.

Nuclear Structure

The chemist's model of the nucleus uses sub-atomic "building blocks," neutrons and protons, to describe the nucleus. Physicists describe neutrons and protons as made up of even more fundamental particles. However, we'll use the simpler model of the nucleus as a collection of "nucleons," protons and neutrons.

Nuclei which each have the same number of protons, but different numbers of neutrons are isotopes. $^{235}_{92}U$ and $^{238}_{92}U$ are different isotopes of uranium.

Problem 21-1:

Complete the following table:

	Symbol	Number of Protons	Number of Neutrons	Atomic Mass Number
a.	$^{13}_{6}C$	–	–	13
b.	–	17	20	–
c.	–	–	19	35

Solution:

For the symbol $^{A}_{Z}X$

> Z is the number of protons in the nucleus.

> A is the number of nucleons (protons + neutrons) in the nucleus.

> X is the symbol for the element of atomic number Z.

a. $^{13}_{6}C$ indicates carbon, with 13 nucleons in the nucleus; 6 of the nucleons are protons; (13 – 6) = 7 of the nucleons are neutrons.

b. This nucleus has 17 protons: the element with atomic number 17 is chlorine, Cl. The nucleus has 17 protons and 20 neutrons, 37 nucleons. The symbol for this isotope of chlorine is $^{37}_{17}Cl$.

c. This nucleus has 35 nucleons, 19 neutrons, and (35 – 19) = 16 protons. The element with atomic number 16 is sulfur, S. The symbol for this nuclide is $^{35}_{16}S$.

Many nuclei behave as though they are spinning charges. Nuclei with an odd number of protons and/or an odd number of neutrons have a net "spin." These nuclei behave like tiny magnets when placed in a magnetic field. Organic chemists study nuclei like ^1_1H and $^{13}_6C$ in magnetic fields to help determine structures of organic molecules.

Nuclear Reactions

We'll look at three sorts of nuclear reactions:

i) Radioactivity: The spontaneous "decay" of unstable nuclei.

ii) Fission: splitting nuclei into new nuclei of similar size.

iii) Fusion: building heavier nuclei from lighter nuclei.

Each of these nuclear reactions is accompanied by large changes in energy.

Balancing Nuclear Equations

Uranium-238 decays to thorium-234, emitting an "alpha particle." An alpha particle is a helium nucleus, ^4_2He. We can write the nuclear equation:

$$^{238}_{92}U \rightarrow ^{234}_{90}Th + ^4_2He$$

We can balance a nuclear equation using the following rules:

i) The total number of nucleons must be the same on both sides of the equation; in this case there are 238 nucleons on the left hand side and (234 + 4) nucleons on the right hand side.

ii) The total charge must be the same on both sides of the equation, in this case there is a positive charge of 92 on the uranium nucleus, 90 on the thorium nucleus, and 2 on the helium nucleus: 92 = (90 + 2).

Problem 21-2:

Balance the following nuclear reactions:

a. $^4_2He + ^{14}_7N \rightarrow ^1_1H + $ ___

b. $^{214}_{84}Po \rightarrow ^4_2He + $ ___

Solution:

Each answer is of the form $^A_Z X$

a. Balance the superscripts:

$$4 + 14 = 1 + A \quad : \quad A = 17$$

301

Balance the subscripts:

$$2 + 7 = 1 + Z \quad : \quad Z = 8$$

Find the symbol in the periodic table with atomic number 8:

$$^{17}_{8}O$$

Therefore: $^{4}_{2}He + ^{14}_{7}N \rightarrow ^{1}_{1}H + ^{17}_{8}O$

b. Balance the superscripts:

$$214 = 4 + A \quad : \quad A = 210$$

Balance the subscripts:

$$84 = 2 + Z \quad : \quad Z = 82$$

$$^{210}_{82}Pb$$

Therefore: $^{214}_{84}Po \rightarrow ^{4}_{2}He + ^{210}_{82}Pb$

Nuclear Stability

Figure 21.1 shows a zone of stability for nuclei.

i) All known nuclei of atomic number greater than 83 are unstable and will decay.

ii) Neutrons seem to be necessary for a nucleus to be stable. Light stable nuclei (except for $^{1}_{1}H$) have at least as many neutrons as protons. Heavier stable nuclei have more neutrons than protons.

iii) Nuclei with an even number of both protons and neutrons seem to be more stable than those with odd numbers of protons and/or neutrons.

iv) There are magic numbers of protons or neutrons which seem to yield stable nuclei.

Radioactive Decay

Radioactive nuclei can decompose in various ways:

i) α emission: helium nuclei are produced; for example

$$^{222}_{86}Rn \rightarrow ^{4}_{2}He + ^{218}_{84}Po$$

The mass number of the nucleus decreases by 4; the atomic number decreases by 2. Heavy nuclei are often α-emitters. α-emission is often accompanied, by emission of high energy photons, γ (gamma) rays.

ii) β emission: β particles, electrons, are emitted by the decaying nucleus. β particles can be denoted as $_{-1}^{0}\beta$ or $_{-1}^{0}e$:

$$_{6}^{14}C \rightarrow {}_{7}^{14}N + {}_{-1}^{0}e$$

The mass number of the nucleus remains the same; the atomic number, the number of protons, <u>increases</u> by one. The number of neutrons <u>decreases</u> by one. We would expect β emission for nuclei which have too many neutrons to be stable.

iii) Positron Emission: A positron is a particle with the same mass as an electron, but a positive charge. We can use the symbols $_{+1}^{0}\beta$ or $_{+1}^{0}e$:

$$_{19}^{38}K \rightarrow {}_{18}^{38}Ar + {}_{+1}^{0}e$$

Again, the mass number remains the same, but the neutron to proton ratio changes.

iv) Electron Capture: An inner orbital electron is captured by the nucleus:

$$_{4}^{7}Be + {}_{-1}^{0}e \text{ (orbital electron)} \rightarrow {}_{3}^{7}Li$$

Check to see that each of these equations is balanced.

Kinetics of Radioactive Decay

The number of nuclei that decay in a given time is proportional to the number of radioactive nuclei present. We can write a rate law for radioactive decay:

$$Rate = -\frac{\Delta N}{\Delta t} = kN$$

where N is the number of radioactive nuclei, and k is the first order rate constant. This is the rate law for a <u>first-order</u> process. The integrated form of this rate law is:

$$\ln\left(\frac{N}{N_o}\right) = -kt \text{ or } N = N_o e^{-kt}$$

where N_o is the number of radioactive nuclei present when t = 0, and N is the number of radioactive nuclei present at time t.

Remember, for a first-order process; the half-life, $t_{1/2}$ is given by:

$$t_{1/2} = \frac{0.693}{k} \text{ or } k = \frac{0.693}{t_{1/2}}$$

Each radioactive nucleus has a characteristic half-life:

$^{235}_{92}$U has a half-life of 7.1×10^6 years;

$^{14}_{6}$C, 5730 years;

$^{32}_{15}$P, 14.28 days;

$^{222}_{86}$Rn, 3.82 days.

Problem 21-3:

If a chemist purchases 5.00 grams of $^{32}_{15}$P, how much remains after 42.84 days?

Solution:

How many half-lives is 42.84 days?

$$\frac{42.84 \text{ days}}{14.28 \text{ days/half-life}} = 3.00 \text{ half-lives;}$$

because 42.84 days is an exact number of half-lives, the problem is simple:

Time	Amount of $^{32}_{15}$P remaining
0	5.00 grams
14.28 days	2.50 grams: 1/2 (5.00 grams)
28.56 days	1.25 grams: 1/2 (2.50 grams)
42.84 days	0.625 grams: 1/2 (1.25 grams)

Problem 21-4:

If a chemist purchases 5.00 grams of $^{32}_{15}$P, how much remains after 9.50 days?

Solution:

9.50 days is less than one half life. We can use the equation:

$$N = N_o e^{-kt}$$

if we can calculate k. Use the expression for half-life:

$$k = \frac{0.693}{t_{1/2}}$$

$$k = \frac{0.693}{14.28 \text{ days}} = 4.85 \times 10^{-2} \text{ days}^{-1}$$

$$N = 5.00 \text{ grams} \times e^{-(4.85 \times 10^{-2} \text{ days}^{-1})(9.50 \text{ days})}$$

$$= 5.00 \text{ grams} \times e^{-.461}$$

$$= 5.00 \text{ grams } (.631) = 3.16 \text{ grams}$$

We often measure the concentration of a radioactive substance by counting the number of "disintegrations" of radioactive nuclei per minute. The disintegrations per minute (dpm) is proportional to the concentration of radioactive nuclei:

$$\text{Rate} = kN$$

We can use the rate of decay of radioactive nuclei to determine the ages of rocks and to date wooden artifacts.

For example:

Carbon-14 is produced in the atmosphere by cosmic rays. Living plants incorporate carbon-14 in carbohydrates produced by photosynthesis. Animals incorporate the carbon-14 by consuming plants, or plant-eating animals. All living plants and animals, then, have the same ratio of carbon-14 to carbon-12 as the atmosphere. When the plant or animal dies no more carbon-14 is incorporated and the concentration of carbon-14 decreases, with a half-life of 5730 years.

Problem 21-5:

The decay rate of $^{14}_{6}C$ in freshly cut wood is 13.6 dpm/gram of carbon. If a piece of charred wood found at an archeological site has a $^{14}_{6}C$ decay rate of 8.50 dpm/gram, how long ago was the charred wood cut from a tree? (We must assume that the concentration of $^{14}_{6}C$ in the wood when it was cut down was the same as the concentration of $^{14}_{6}C$ in the live trees now. This assumption has been found to be a simplification.

Solution:

Use the equation:

$$N = N_o e^{-kt}$$

$$\text{or} \quad \ln\left(\frac{N}{N_o}\right) = -kt$$

First we need k:

$$k = \frac{0.693}{t_{1/2}} = \frac{0.693}{5730 \text{ years}}$$

$$= 1.21 \times 10^{-4} \text{ years}^{-1}$$

The dpm are proportional to the concentration of $^{14}_{6}C$:

$$\frac{N}{N_o} = \frac{8.50}{13.6} = 0.625$$

$$-kt = \ln\left(\frac{N}{N_o}\right) = \ln(0.625)$$

$$-kt = -.470$$

$$t = \frac{.470}{k} = \frac{.470}{1.21 \times 10^{-4} \text{ years}^{-1}} = 3.88 \times 10^3 \text{ years}$$

Binding Energy

One $^{14}_{7}$N atom weighs 14.00307 amu. This atom contains 7 electrons, each weighing 0.000549 amu. The $^{14}_{7}$N nucleus must weigh:

$$\begin{array}{ll} 14.00307 \text{ amu} & ^{14}_{7}\text{N atom} \\ -7(.000549 \text{ amu}) & 7 \text{ electrons} \\ \hline 13.99923 \text{ amu} & ^{14}_{7}\text{N nucleus} \end{array}$$

If we "built" this nucleus from 7 neutrons, each weighing 1.00866 amu, and 7 protons, each weighing 1.00728 amu, we would expect a mass of:

$$\begin{array}{lll} 7 \text{ neutrons:} & 7(1.00866 \text{ amu}) = & 7.06062 \text{ amu} \\ 7 \text{ protons:} & 7(1.00728 \text{ amu}) = & 7.05096 \text{ amu} \\ & & \hline \\ & & 14.11158 \text{ amu} \end{array}$$

Some mass seems to be "missing" from the nucleus:

$$\begin{array}{l} 14.11158 \text{ amu} \\ -13.99923 \text{ amu} \\ \hline .11235 \text{ amu} \end{array}$$

If one mole of nitrogen nuclei were formed from seven moles of neutrons and seven moles of protons, 0.11235 grams of mass would be "missing." The missing mass was converted to energy when the nuclei were formed. If we wished to break the nitrogen nuclei apart we would need to provide this energy, the "binding energy."

We can calculate the binding energy from the "missing mass" (or "mass defect") using Einstein's Equation:

$$\Delta E = \Delta mc^2$$

ΔE is the change in energy accompanying

Δm, a change in mass

c is the speed of light

Problem 21-6:

How much energy is given off when one mole of nitrogen-14 nuclei is formed from 7 moles of neutrons and 7 moles of protons?

Solution:

The "missing" mass is: 0.11235 g/mole = 1.1235×10^{-4} kg/mole

$$\Delta E = \Delta m \cdot c^2$$

$$= 1.1235 \times 10^{-4} \text{ kg/mole } (2.9979 \times 10^8 \text{ m/s})^2$$

$$= 1.1235 \times 10^{-4} \text{ kg/mole } (8.9874 \times 10^{16} \text{ m}^2/\text{s}^2)$$

$$= 1.0097 \times 10^{13} \text{ kg m}^2/\text{s}^2/\text{mole} = 1.0097 \times 10^{13} \text{J/mole}$$

$$= 1.0097 \times 10^{10} \text{ kJ/mole}$$

To break one mole of nitrogen nuclei into protons and neutrons would require 1.0097×10^{10} kJ.

The energy to break a mole of N_2 bonds, on the other hand, is 941 kJ. The binding energy of a nitrogen nucleus is more than 10 million times the bond energy of a nitrogen-nitrogen triple bond!

Problem 21-7:

How much mass is "lost" when one mole of N_2 is formed from two moles of nitrogen atoms, if 941 kJ are given off?

Solution:

Use $\Delta E = \Delta m \cdot c^2$

$$\Delta m = \frac{\Delta E}{c^2} = \frac{941 \times 10^3 \text{ J}}{(2.9979 \times 10^8 \text{ m/s})^2}$$

$$= 1.05 \times 10^{-11} \text{ kg}$$

$$= 1.05 \times 10^{-8} \text{ g}$$

This mass is too small to be noticed in the laboratory.

Binding Energy Per Nucleon

If 1.0097×10^{10} kJ are released when a mole of nitrogen nuclei is formed from neutrons and protons, how much energy is released when one nucleus of nitrogen is formed?

$$\text{Binding energy} = \frac{1.0097 \times 10^{10} \text{ kJ}}{6.0221 \times 10^{23} \text{ nuclei}} = \frac{1.0097 \times 10^{13} \text{ J}}{6.0221 \times 10^{23} \text{ nuclei}}$$

$$= 1.6766 \times 10^{-11} \text{ J/nuclei}$$

We can use a more convenient energy unit, MeV (Million electron Volts):

$$1 \text{ MeV} = 1.60219 \times 10^{-13} \text{ J}$$

Therefore, the binding energy for the nitrogen nucleus is:

$$\text{Binding energy} = \frac{1.6766 \times 10^{-11} \text{ J}}{\text{nuclei}} \times \frac{1 \text{ MeV}}{1.60219 \times 10^{-13} \text{ J}}$$

$$= 104.65 \text{ MeV}$$

We can calculate "binding-energy-per-nucleon" so that we can compare the relative stability of different nuclei:

for $^{14}_{7}$N, the binding energy per nucleon is:

$$\frac{104.65 \text{ MeV}}{14 \text{ nucleons}} = \frac{7.48 \text{ MeV}}{\text{nucleon}}$$

Figure 21.10 shows the binding energy per nucleon as a function of mass number.

$^{56}_{26}$Fe is the most stable nucleus, with the largest binding energy per nucleon. Lighter nuclei can be "fused" to give off energy; heavier nuclei can be "fragmented" to give off energy.

CHAPTER TWENTY-ONE: SELF-TEST

21-1. Complete the following table.

Symbol	Number of Protons	Number of Neutrons	Atomic Mass Number
$^{17}_{8}O$	___	___	___
___	___	10	19
___	53	___	131

21-2. Balance the following nuclear equations

a. $^{11}_{6}C \rightarrow {}^{4}_{2}He +$ _____

b. $^{17}_{36}Kr \rightarrow {}^{0}_{-1}\beta +$ _____

c. $^{238}_{92}U + {}^{1}_{0}n \rightarrow$ _____ $+ {}^{0}_{-1}\beta$

d. $^{236}_{92}U \rightarrow {}^{91}_{36}Kr + 3{}^{1}_{0}n +$ ____

21-3. Decide if each statement is true or false:

i) Nuclei with too many neutrons per proton tend to undergo positron ($^{0}_{1}\beta$) emission.

ii) Nuclei with an even number of protons and even number of neutrons tend to be stable.

iii) "γ rays" are high energy photons.

iv) The half-life for decay of a radioactive nucleus depends on the compound the nucleus is found in.

v) Stable heavy nuclei tend to have the same number of neutrons as protons.

vi) Nuclei with too few neutrons per proton tend to undergo positron ($^{0}_{1}\beta$) emission.

21-4. $^{60}_{27}Co$ undergoes decay by $^{0}_{-1}\beta$ emission, with a half-life of 5.27 years.

a. If a chemist has a 400.0 mg sample of $^{60}_{27}Co$, how many milligrams remain after 15.81 years?

b. How many milligrams remain after 8.28 years?

21-5. The decay rate of $^{14}_{6}C$ in freshly cut wood is 13.6 dpm/gram of carbon. If a piece of a wooden bowl at an archaeological dig has a $^{14}_{6}C$ decay rate of 5.8 dpm/gram, how old is the bowl? $t_{1/2}$ for $^{14}_{6}C$ is 5730 years.

21-6.   The <u>atomic</u> mass of $^{7}_{3}$Li is 7.016005 amu.

   a.   The mass of an electron is 0.000549 amu. What is the mass of a $^{7}_{3}$Li <u>nucleus</u>?

   b.   The mass of one neutron is 1.00866 amu, and the mass of one proton is 1.00728 amu. What is the mass of 4 neutrons and 3 protons?

   c.   What mass appears to be "missing" from the $^{7}_{3}$Li nucleus?

   d.   What mass would be "missing" from a mole of $^{7}_{3}$Li nuclei?

   e.   How much energy would be given off when one <u>mole</u> of $^{7}_{3}$Li nuclei was formed from neutrons and protons?

21-7.   For the nuclear reaction

$$^{226}_{88}\text{Ra} \rightarrow \, ^{222}_{86}\text{Rn} + \, ^{4}_{2}\text{He}$$

There is 0.0052 g "lost" per mole of Ra that reacts. Do nuclear reactions violate the law of conservation of mass?

CHAPTER TWENTY-ONE: SELF-TEST ANSWERS

21-1.

Symbol	Number of Protons	Number of Neutrons	Atomic Mass Number
$^{17}_{8}O$	8	9	17
$^{19}_{9}F$	9	10	19
$^{131}_{53}I$	53	78	131

21-2 a. $^{11}_{6}C \rightarrow {}^{4}_{2}He + {}^{7}_{4}Be$

b. $^{87}_{36}Kr \rightarrow {}^{0}_{-1}\beta + {}^{87}_{37}Rb$

c. $^{238}_{92}U + {}^{1}_{0}n \rightarrow {}^{239}_{93}Np + {}^{0}_{-1}\beta$

d. $^{236}_{92}U \rightarrow {}^{91}_{36}Kr + 3\ {}^{1}_{0}n + {}^{142}_{56}Ba$

21-3. i) false

ii) true

iii) true

iv) false

v) false

vi) true

21-4. a. After 3 half-lives 50 mg $^{60}_{27}Co$ remain

b. $\ln \frac{N_o}{N} = kt$; $k = \frac{0.693}{t_{1/2}} = 0.1315\ hr^{-1}$

$$\ln \frac{N_o}{N} = 0.1315\ yr^{-1}\ (8.28\ yr) = 1.09$$

$$\frac{N_o}{N} = 2.97$$

$$N = 135\ mg$$

21-5. $-kt = \ln \left(\frac{N}{N_o}\right)$; $k = \frac{0.693}{t_{1/2}} = 1.21 \times 10^{-4}\ yrs^{-1}$

$$t = \frac{-1}{k} \ln \left(\frac{N}{N_o}\right) = \frac{-1}{1.21 \times 10^{-4}\ yr^{-1}} \ln\left[\frac{5.8}{13.6}\right] = 7.0 \times 10^{3}\ years$$

21-6.   a.   7.016005 amu − 3(0.000549 amu) = 7.014853 amu

b.   4(1.00866 amu) + 3(1.00728 amu) = 7.05648 amu

c.   7.05648 amu − 7.014853 amu = 0.04163 amu

d.   0.04163 g = 4.163×10^{-5} kg

e. $\Delta E = \Delta mc^2$

$$= (4.163 \times 10^{-5} \text{ kg})(2.9979 \times 10^8 \text{ m/s})^2 = 3.741 \times 10^{12} \text{ J/mole}$$

$$= 3.741 \times 10^9 \text{ kJ/mole}$$

21-7.   The mass "lost" is converted to energy. Matter is not conserved separately from energy. Mass can be converted to energy; energy can be converted to mass.

CHAPTER TWENTY-TWO: ORGANIC CHEMISTRY

This chapter offers an introductory overview of the chemistry of carbon compounds. Carbon has four valence electrons and can form four covalent bonds. (Rarely is a carbon atom found with a lone pair of electrons, as in carbon monoxide, :C O:). Carbon can use sp^3 hybrids to form 4 σ bonds. Carbon can use sp^2 hybrids to form 3 σ bonds and a p orbital to form a π bond. Carbon can use sp hybrids to form 2 σ bonds, and 2 p orbitals to form 2 π bonds, or one triple bond. Carbon can bond to other carbon atoms to form the backbone of long chains, branched networks of atoms, or rings.

Problem 22-1:

Describe the atomic or hybrid orbitals used to form each bond in the following molecules. How large is each of the angles indicated by arrows?

Solution:

First, determine the hybrids used by each carbon atom: i) count the p orbitals needed for π bonds and use the remaining p orbitals to form hybrids; or, ii) count the σ bonds formed by the carbon: form one hybrid for each σ bond and lone pair. Then, predict the bond angles from VSEPR (these angles could be used to predict the hybrids). (Carbon rarely has lone pairs.)

a. Each carbon atom forms 4 σ bonds: therefore, each uses sp^3 hybrids. Each hydrogen atom uses 1s orbitals. Each bond angle is 109.5°.

b. Each carbon atom forms 3 σ bonds and 1 π bond; therefore, each uses sp^2 hybrids to form the σ bonds, and a p orbital to form the π bond. Each bond angle is 120°.

c. The two "left-most" carbons each form 2 σ bonds and 2 π bonds; therefore, each uses sp hybrids to form the σ bonds, and 2p orbitals to form the two π bonds. The "right hand" carbon forms 4 σ bonds, using sp^3 hybrids. The bond angles around the "left-most" carbons are 180°. The bond angles around the "right hand" carbon are 109.5°.

Although we draw these molecules as though they are "flat" (planar), most are not. If you build the molecules in Problem 22-1, you'll find that only b. is planar.

We often write structures of organic molecules in "condensed form." a. above becomes CH_3-CH_3; b. becomes $CH_2=CH_2$; and c. becomes $HC\equiv C-CH_3$. You are expected to be able to "translate" these "condensed" structural formulas into the Lewis structures.

Organic compounds are classified in groups with similar chemistry. The first class of organic compounds we look at are the hydrocarbons, compounds containing only carbon and hydrogen. Hydrocarbons which contain no double or triple bonds are alkanes.

$$\begin{array}{ccc}
\begin{array}{c}
H \\
| \\
H - C - H \\
| \\
H
\end{array}
&
\begin{array}{c}
H \quad H \\
| \quad | \\
H - C - C - H \\
| \quad | \\
H \quad H
\end{array}
&
CH_3 - CH_2 - CH_3
\end{array}$$

CH_4	CH_3-CH_3	
Methane	Ethane	Propane

Alkanes containing four or more carbon atoms can be branched. For example, there are two isomers of C_4H_{10}:

$$CH_3 - CH_2 - CH_2 - CH_3 \qquad\qquad
\begin{array}{c}
CH_3 \\
| \\
CH_3 - CH - CH_3
\end{array}$$

Butane	Methyl Propane
(n–Butane)	(iso–Butane)

In methylpropane, a methyl group has been "substituted" for a hydrogen atom on the central carbon in propane.

Naming Alkanes

1. Find the longest chain (with the largest number of substituents).

2. Name the longest chain using Greek roots for chains longer than four carbons:

1 carbon	meth–		5 carbons	pent–
2 carbons	eth–		6 carbons	hex–
3 carbons	prop–		7 carbons	sept–
4 carbons	but–		8 carbons	oct–
			9 carbons	dec–

3. Name the substituents (side groups) on the chain.

4. Number the carbon atoms in the longest chain, starting from the end closest to a substituent.

5. Name the compound, giving the address for each substituent on the chain. Name the substituents in alphabetical order. Use di–, tri–, etc. if the same substituent occurs more than once.

314

Problem 22-2:

Name the following compounds:

a. 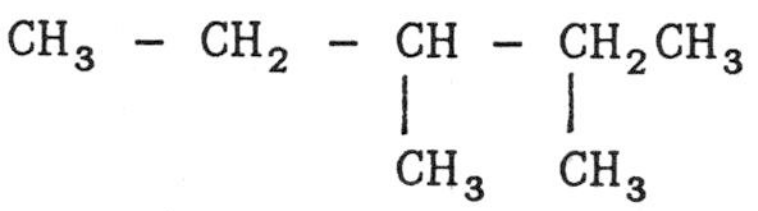 $CH_3 - CH_2 - CH - CH_2CH_3$
               | |
             CH_3 CH_3

b. 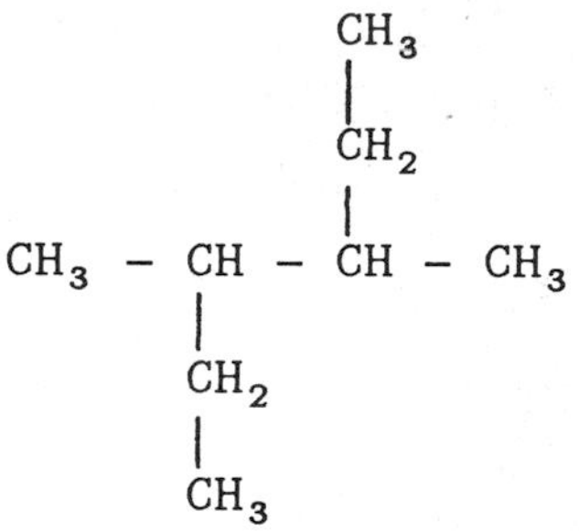 $CH_3 - CH - CH - CH_3$
with substituents:
CH_3
|
CH_2
|
on the CH: CH_2
|
CH_3

Solution:

a. The longest chain has 5 carbon atoms. Number the chain from the right to give the lowest "addresses." 2,3-dimethylpentane.

b. The longest chain has 6 carbon atoms. We can number the chain from either end. 3,4-dimethylhexane (not 2,3-diethylbutane!).

Problem 22-3:

Write the structural formulae for the following compounds:

a. 3-ethyl-2-methyl hexane

b. 2,3,4-trimethylpentane

Solution:

Draw and number the longest chain. Then add the substituents, and add hydrogen atoms to complete the structure.

a. i) Draw and number the six carbon chain:

$$C - C - C - C - C - C$$
$$1 \quad 2 \quad 3 \quad 4 \quad 5 \quad 6$$

ii) Add the substituents (side groups):

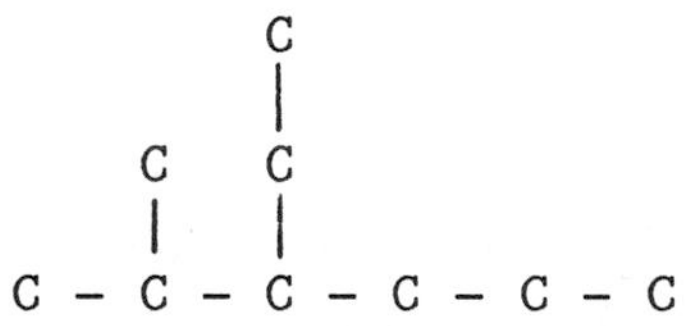

```
            C
            |
        C   C
        |   |
    C - C - C - C - C - C
```

iii) Complete the structure with hydrogen atoms:

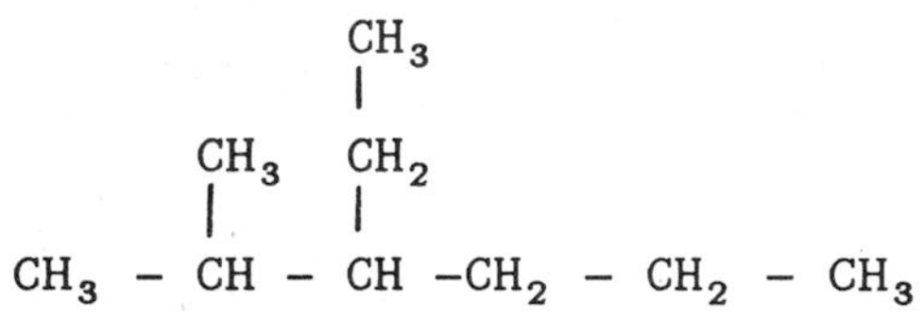

```
                CH3
                |
        CH3     CH2
        |       |
    CH3 - CH - CH -CH2 - CH2 - CH3
```

315

$$
\begin{array}{ccccc}
& \text{CH}_3 & \text{CH}_3 & \text{CH}_3 & \\
& | & | & | & \\
\text{b.} \qquad \text{CH}_3 - & \text{CH} - & \text{CH} - & \text{CH} - & \text{CH}_3
\end{array}
$$

Reactions of Alkanes

Alkanes are relatively unreactive. At high temperatures alkanes "burn" in oxygen to form carbon dioxide and water.

Problem 22-4:

Balance the reaction for the combustion of hexane with oxygen.

Solution:

i) First write the formulae of the reactants and products:

$$\underline{\quad} \; C_6H_{14} + \underline{\quad} \; O_2 \rightarrow \underline{\quad} \; CO_2 + \underline{\quad} \; H_2O$$

ii) Balance for carbon and hydrogen:

$$1 \; C_6H_{14} + \underline{\quad} \; O_2 \rightarrow 6 \; CO_2 + 7 \; H_2O$$

iii) Then balance for oxygen: there are 19 oxygen atoms on the right hand side:

$$C_6H_{14} + \frac{19}{2} \; O_2 \rightarrow 6 \; CO_2 + 7 \; H_2O$$

or: $\qquad 2 \; C_6H_{14} + 19 \; O_2 \rightarrow 12 \; CO_2 + 14 \; H_2O$

Now you can use the balanced equation to do stoichiometry calculations!

In the presence of light, or at high temperatures, alkanes undergo substitution reactions with halogens. For example:

$$CH_3CH_3 + Cl_2 \xrightarrow{h\nu} CH_3CH_2Cl + HCl$$

The product, chloroethane, can react further with chlorine to form di-, tri-, etc. chloroethanes.

Cycloalkanes

Cycloalkanes are alkanes containing rings. 3-membered rings (cyclopropane) are very strained: the bond angles are 60°. Cyclobutane and cyclopentane are less stained. Cyclohexane is very stable. These rings are not flat, although we often draw them as though they were. Cyclobutane is "puckered," a square which is slightly folded along the diagonal. Cyclohexane has two "conformations," the boat and the chair. If you build a model of cyclohexane you'll see why the chair conformation is more stable.

Unsaturated Hydrocarbons

Hydrocarbons with carbon–carbon double bonds are alkenes. The double bond is a π bond formed by side-to-side overlap of p orbitals. There is no rotation about the double bond.

Hydrocarbons with carbon–carbon triple bonds are alkynes. To name alkenes and alkynes, name the longest carbon–carbon chain containing the double or triple bond. Number the chain from the end closest to the multiple bond. The "address" of the double bond is that of the lower numbered carbon.

$$CH_3 - C \!=\! C - CH_3$$

with CH_3 above the left C and $CH_2 - CH_3$ below the right C.

$$
\begin{array}{c}
\qquad\quad CH_3 \\
\qquad\quad | \\
CH_3 - C \!=\! C - CH_3 \\
\qquad\quad | \\
\qquad\quad CH_2 \\
\qquad\quad | \\
\qquad\quad CH_3
\end{array}
$$

The longest chain contains 5 carbon atoms. Number the chain from the left. The backbone, then, is 2-pentene. The molecule is 2,3-dimethyl-2-pentene.

Reaction of Alkenes and Alkynes

Unsaturated molecules undergo addition reactions, where π bonds are broken and new σ bonds form. These reactions include hydrogenation, the addition of hydrogen, and halogenation, the addition of a halogen.

Hydrogenation: $CH_3 - CH \!=\! CH - CH_3 + H_2 \xrightarrow{Pt} CH_3 - CH_2 - CH_2 - CH_3$

where platinum is the catalyst.

Halogenation: $CH_3 - CH \!=\! CH - CH_3 + Cl_2 \longrightarrow CH_3 - \underset{\underset{Cl}{|}}{CH} - \underset{\underset{Cl}{|}}{CH} - CH_3$

Aromatic Hydrocarbons

The π electrons in the aromatic ring in benzene are delocalized around the ring. The aromatic ring is very stable, undergoing substitution reactions, not addition reactions.

$$\bigcirc + Br_2 \xrightarrow{FeBr_3} \bigcirc\!-\!Br + HBr$$

Hydrocarbon Derivatives

We can view many organic molecules as derivatives of hydrocarbons, many with a functional group "substituted" for a hydrogen atom, or hydrogen atoms in the original molecule. The chemistry of these compounds is characteristic of the functional group. Table 22.3 lists some common functional groups.

How can we organize our study of these molecules? We need to consider

    i)   Nomenclature

    ii) Physical properties, especially boiling point, solubility

    iii) Preparation

and iv) Reactions

Nomenclature

We won't attempt to name very complex molecules containing many functional groups, but here are some general principles:

1) Name the longest continuous chain that contains the functional group. If a molecule contains more than one functional group, choose the "highest priority" functional group. For example, a molecule containing an $-$ OH group and a $-$ COOH group will always be named as an acid, not as an alcohol.

2) Number the chain from the end closest to the principal functional group.

$$\overset{\displaystyle O}{\overset{\displaystyle \|}{HO - CH_2 - CH_2 - CH_2 - COH}}$$

is 4-hydroxybutanoic (or butyric) acid.

Boiling Point

The boiling point of a liquid depends on the interactions between the molecules. The attractions between non-polar molecules are weak London dispersion forces, due to transient dipoles. London dispersion forces tend to increase with increasing molecular weight. There are stronger dipole-dipole forces between polar molecules. Other molecules can form hydrogen bonds like those found in water.

Problem 22-5:

In each pair below, choose the higher boiling compound.

a. $CH_3CH_2CH_3$ (MW 44) CH_3CH_2OH (MW 46)

b. CH_3-O-CH_3 (MW 46) $CH_3CH_2CH_3$ (MW 44)

c. CH_3CH_3OH (MW 46) CH_3-O-CH_3 (MW 46)

Solution:

   a. Alcohols, compounds containing the − OH functional group, can hydrogen bond. Alcohols are higher boiling than hydrocarbons of the same molecular weight.

b, c. Ethers, molecules containing a C − O − C bridge are slightly polar. Ethers tend to be slightly higher boiling than hydrocarbons of the same molecular weight, lower boiling than alcohols of the same molecular weight.

Compound	bp (°C)
CH_3CH_2OH	78
CH_3-O-CH_3	−24
$CH_3CH_2CH_3$	−42

Solubility

Non-polar compounds will be very soluble in non-polar solvents and insoluble in water. Polar compounds and those which can hydrogen bond with water will be more soluble in water. Remember our "step-wise" view of the process of dissolving:

   i) "Break" the interactions between solute molecules;

   ii) "Break" the interactions between solvent molecules to create "holes" for the solute molecules.

   iii) Place the solute molecules in the "holes" and let the solute and solvent molecules interact.

Problem 22-6:

In each set below, choose the compound which is most soluble in water and the compound which is least solute in water.

   a. $CH_3-CH_2-CH_3$ CH_3-CH_2-OH CH_3-O-CH_3

       Propane   Ethanol   Dimethyl Ether

   b. $CH_3-CH_2-CH_2-OH$ $CH_3-CH_2-CH_2-CH_2-OH$ $CH_3-CH_2-CH_2-CH_2-CH_2-OH$

       1-Propanol   1-Butanol   1-Pentanol

   c. $CH_3CH_2CH_2CH_2CH_2CH_3$ $HO-CH_2-CH_2-CH_2-CH_2-OH$ $CH_3-CH_2-CH_2-CH_2-CH_2-OH$

       Hexane   1,4-Butanediol   1-Pentanol

Solution:

a. Ethanol is the most soluble in water. The oxygen in ethanol can form hydrogen bonds with water using its two lone pairs of electrons, and using the hydrogen atom bound to the oxygen. Dimethyl ether can hydrogen bond to water using the two lone pairs of electrons on the oxygen atom. Propane, a non-polar hydrocarbon, is not soluble in water.

b. Each of these alcohols can hydrogen bond to water. However, the longer the hydrocarbon chain, the larger "hole" we need to create in the water. 1-propanol is the most soluble; 1-pentanol is the least soluble.

c. 1,4-butanediol is the most soluble; it has two sites to use in hydrogen bonding with water. Hexane is the least soluble.

Preparation and Reactions

Try to write a "generic" reaction for each process, so that you can focus on the chemistry of the functional group.

For example:

"An alcohol may be converted to an alkene by dehydration in the presence of an acid catalyst" can be written:

$$- CH_2 - \underset{\underset{OH}{|}}{CH} - \xrightarrow[\text{heat}]{85\% \ H_3PO_4} - HC = CH - + H_2O$$

Or:

"A primary alcohol can be oxidized by potassium permanganate to give a carboxylic acid" can be written as:

$$- CH_2 - \ddot{O}H \xrightarrow{KMnO_4} - \overset{\overset{\ddot{O}}{||}}{C} - \ddot{O}H$$

Now you can apply these "generic reactions" to particular problems.

Problem 22-7:

Complete the following reactions:

a. $CH_3-CH_2-CH_2-CH_2-OH \xrightarrow{KMnO_4}$

b. $CH_3-CH_2-CH_2-CH_2-OH \xrightarrow{85\% \ H_3PO_4}$

Solution:

a. Use our "generic" reaction for oxidation of primary alcohols.

$$- CH_2OH \xrightarrow{KMnO_4} - \overset{\overset{\displaystyle O}{\parallel}}{C}OH$$

Then:
$$CH_3-CH_2-CH_2-CH_2OH \xrightarrow{KMnO_4} CH_3-CH_2-CH_2-\overset{\overset{\displaystyle O}{\parallel}}{C}-OH$$

b. Use our "generic" reaction for dehydration of alcohols:

$$- CH_2-\overset{\overset{\displaystyle OH}{|}}{C}H - \xrightarrow[\text{heat}]{85\% \ H_3PO_4} - CH = CH -$$

Then: the reaction of $CH_3CH_2CH_2-CH_2OH$ can be written

$$CH_3CH_2CH_2 - \overset{\overset{\displaystyle OH}{|}}{\underset{\underset{\displaystyle H}{|}}{C}} - H \xrightarrow[\text{heat}]{85\% \ H_3PO_4} CH_3CH_2CH = CH_2$$

The study of organic chemistry does require memorization of reactions. The task is made easier when we find similarities in mechanisms of reactions, when we understand "how" and "why" a reaction occurs.

Try writing "generic" reactions for the organic reactions in Chapter 22.

Remember, although we "treat" organic chemistry as a separate subject, we can still rely on the same principles of bonding, kinetics, thermodynamics, equilibria and stoichiometry that we've already developed.

22-1. Describe the hybrid orbitals used by each carbon atom in the following molecules: (Hint: Draw the Lewis structure.)

a. CH$_3$-CH$_2$-OH

c. CH$_3$COH with =O above C

b. HC≡CH

d.

22-2. Draw and name all the isomers of C$_5$H$_{12}$.

22-3. Draw all the isomers of C$_4$H$_8$.

22-4. Draw the structural formula for 3,5-diethyl-2-methyl-4-propylheptane.

22-5. Balance the following reactions.

a. ___ CH$_3$CH$_2$OH + ___ O$_2$ → ___ CO$_2$ + ___ H$_2$O

b. ___ CH$_3$CH$_2$OH + ___ MnO$_4^-$ → ___ Mn^{2+} + ___ CH$_3$C(O)OH   in acidic solution

22-6. Name the following compounds.

a. H$_2$C=CH-CH$_3$

b. CH$_3$-C=C-CH$_3$

22-7. Match each formula with its class.

a. CH$_3$NH$_2$

b. CH$_3$COH (with O above)

c. CH$_3$-O-CH$_3$

d. CH$_3$CH$_2$OH

e. CH$_3$-C CH$_2$ (with H above)

f. CH$_3$CCH$_3$ (with O above)

g. CH$_3$CH$_2$CH$_2$CH$_3$

h. HC C-CH$_2$-CH$_3$

i) alcohol

ii) ketone

iii) ether

iv) carboxylic acid

v) alkene

vi) alkyne

vii) alkane

viii) amine

22-8. Complete the following equations

a. $$CH_3\overset{\overset{\textstyle O}{\textstyle \|}}{C}OH + OH^- \rightarrow$$

b. Dehydration of an alcohol

$$CH_3-\underset{\overset{|}{OH}}{CH}-CH_3 \xrightarrow{\ H_2SO_4\ }$$

c. Addition of a halogen to an alkene

$$CH_2\ CH_2-CH_3 \xrightarrow{\ Br_2\ }$$

d. Addition of hydrogen to an alkene

$$CH_2\ CH-CH_3 \xrightarrow{\ H_2,\ Pt\ }$$

22-1. a. Each carbon uses sp^3 hybrids.

b. Each carbon uses sp hybrids.

c. The carbon on the left uses sp^3 hybrids; the carbon on the right uses sp^2 hybrids.

d. The center carbon uses sp hybrids; the end carbons use sp^2 hybrids.

22-2. $CH_3CH_2CH_2CH_2CH_3$ n-pentane

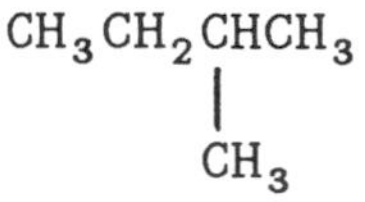

$CH_3CH_2CHCH_3$ 2-methylbutane
           |
          CH_3

              CH_3
              |
CH_3-C-CH_3 1,2-dimethylpropane
              |
              CH_3

22-3. $CH_2=CH-CH_2-CH_3$

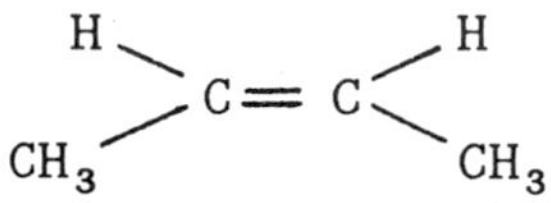

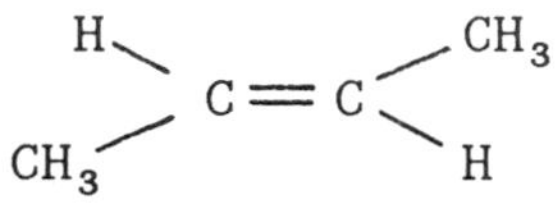

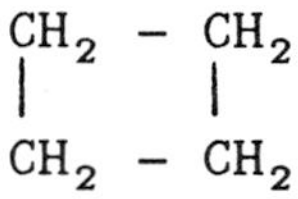

$CH_2 - CH_2$
| |
$CH_2 - CH_2$

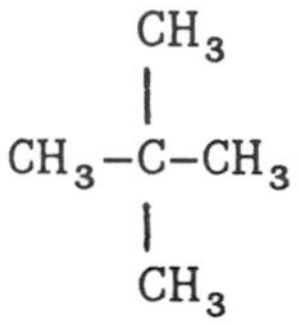

22-4. CH_3
             |
$CH_3-CH-CH--CH--CH-CH-CH_3$
         | | |
        CH_2 CH_2 CH_2
         | | |
        CH_3 CH_2 CH_3
              |
              CH_3

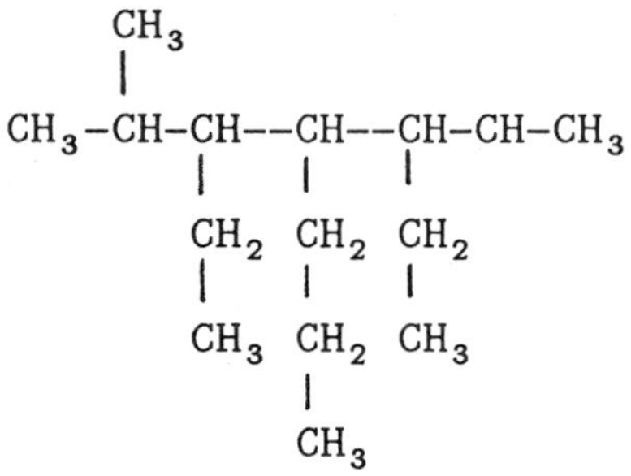

22-5. a. $CH_3CH_2OH + 3\ O_2 \rightarrow 2\ CO_2 + 3\ H_2O$

b. $5\ CH_3CH_2OH + 12\ H^+ + 4\ MnO_4^- \rightarrow 5\ CH_3\overset{\overset{\displaystyle O}{\|}}{C}OH + 4\ Mn^{2+} + 11\ H_2O$

22-6. a. propene

b. 2-butyne

22-7. a - viii

   b - iv

   c - iii

   d - i

   e - v

   f - ii

   g - vii

   h - vi

$$
22\text{-}8.\quad \text{a.} \quad \rightarrow CH_3\overset{\overset{\textstyle O}{\|}}{C}O^- + H_2O
$$

   b. $\rightarrow CH_3 - C = CH_2 + H_2O$

   c. $\rightarrow \overset{\overset{\textstyle Br}{|}}{CH_2} - \overset{\overset{\textstyle Br}{|}}{CH} - CH_3$

   d. $\rightarrow CH_3CH_2CH_3$

Chapter 23 gives an overview of the kinds of molecules we encounter in living systems.

Much of biochemistry involves polymers: proteins are polymers of amino acids; complex carbohydrates are polymers of simple sugars; nucleic acids are polymers of "nucleotides" (see Figures 23.23 and 23.24).

Proteins

A protein is a long sequence of amino acids, each linked with a "peptide bond." Each amino acid has the general formula

$$H_2N - \underset{\underset{R}{|}}{\overset{\overset{H}{|}}{C}} - C\overset{\displaystyle O}{\underset{\displaystyle O-H}{\diagup}} \qquad \text{which, in solution, forms the ion} \quad H_3^+N - \underset{\underset{R}{|}}{\overset{\overset{H}{|}}{C}} - C\overset{\displaystyle O}{\underset{\displaystyle O-}{\diagup}}$$

R is the "side group." The simplest amino acid is glycine, with a hydrogen atom as the "side group." All the other naturally occurring amino acids have four different groups bound to the central carbon atom. These amino acids are optically active. Figure 23.3 shows the 20 common amino acids found in proteins. There are non-polar side groups, like

$$CH_3 - \underset{\underset{|}{|}}{\overset{\overset{H}{|}}{C}} - CH_3 \quad \text{in valine;}$$

uncharged polar side groups, like $- CH_2OH$, in sterine; acidic side groups, like

$$- CH_2C\overset{\displaystyle O}{\underset{\displaystyle OH}{\diagup}} \quad \text{, in aspartic acids;}$$

and basic side groups, like $- CH_2CH_2CH_2CH_2NH_2$, in lysine.

Two amino acids can react to form a dipeptide:

$$H_2N - \underset{\underset{R_1}{|}}{\overset{\overset{H}{|}}{C}} - C\overset{\displaystyle O}{\underset{\displaystyle OH}{\diagup}} \quad + \quad H_2N - \underset{\underset{R_2}{|}}{\overset{\overset{H}{|}}{C}} - C\overset{\displaystyle O}{\underset{\displaystyle OH}{\diagup}} \quad \longrightarrow$$

$$H_2N - \overset{\overset{\displaystyle H}{\displaystyle |}}{\underset{\underset{\displaystyle R_1}{\displaystyle |}}{C}} - C\overset{\displaystyle O}{\underset{\underset{\underset{\displaystyle H}{\displaystyle |}}{N}}{\diagup}} - \overset{\overset{\displaystyle H}{\displaystyle |}}{\underset{\underset{\displaystyle R_2}{\displaystyle |}}{C}} - C\overset{\displaystyle O}{\underset{\displaystyle OH}{\diagup}} + H_2O$$

This dipeptide can react with other amino acids to form a polypeptide, a long chain of amino acids. Each protein is a polypeptide with a unique series of amino acids.

The primary structure of a protein is the sequence of amino acids in the polypeptide chain. Many proteins have regions in which the polypeptide chain is coiled into a helix, with hydrogen bonds between the carbonyl oxygen in one turn of the helix and the nitrogen in the peptide link of another turn of the helix, C O ···· H - N . This coiled helix is called the secondary structure of the protein. The side groups point out from the helix. The coiled helix can be bent and folded, as in globular proteins like myoglobin (See Figure 23.9). The polar and ionic side groups are found primarily on the outside of the folded protein, where they can interact with water. The non-polar side groups are found in the interior of the folded protein.

The folding of the coiled protein is called its tertiary structure. The tertiary structure seems to maximize the interaction of polar and ionic side groups with water, and to minimize the interaction of the non-polar side groups with water. The tertiary structure can be stabilized by covalent disulfide bonds, - S - S - links between two cysteine side groups.

The function of a protein depends upon its three dimensional shape. We can disrupt the interactions which lead to the secondary and tertiary structure by changing the pH, heating the solution, adding concentrated salt solutions, using detergents, or using organic solvents. Then we destroy the "biological activity" of the protein.

Carbohydrates

Complex carbohydrates are polymers of "simple sugars." Simple sugars, such as glucose, form rings with an internal "condensation" reaction:

Schematically

Although the glucose ring is actually puckered like a cyclohexane ring.

Figure 23.19 shows how sucrose can be formed from glucose and fructose. Figure 23.20 shows the complex carbohydrates amylose and cellulose. The links in amylose and cellulose are slightly different. Humans have enzymes to hydrolyze the links in amylose, but not the links in cellulose.

Nucleic Acids

Each nucleotide, the building blocks of nucleic acids, contains a purine or pyrimide heterocyclic (containing rings of carbon and nitrogen atoms) base linked to a 5-carbon sugar (a pentose) which is linked to a phosphate by an ester linkage. These nucleotides are linked to form the polymer, nucleic acid, with a sugar-phosphate backbone.

Deoxyribonucleic acid occurs as a double helix, with the two strands of the helix held together by hydrogen bonds between pairs of the bases. Thymine can hydrogen bond to adenine; cytosine can hydrogen bond to guanine. Each strand of DNA is the "complement" of the other. If the sequence of bases on one strand is: - A - T - A - C - C - A - then the sequence on the other is: - T - A - T - G - G - T -, with each base on one strand hydrogen bonding to its complement on the other strand.

Three bases in a sequence in DNA "code" for a single amino acid in a protein. This "message" is "transcribed" to a corresponding ribonucleic acid, messenger RNA. Other small RNA fragments, transfer RNA's, link to particular amino acids, "bring" the amino acids to the messenger RNA so that the protein can be synthesized with a particular sequence of amino acids. The genetic formation contained in the DNA is "transcribed" into the messenger RNA. The message is then "transcribed" into the sequence of amino acids in the protein. Each of these steps depends on molecules "recognizing" each other through hydrogen bonds and ionic interactions.

Lipids

Lipids are naturally occurring organic molecules which are not soluble in water, but are soluble in non-polar solvents.

The simplest lipids are fats, esters of fatty acids with alcohols. The most common fats and oils are triglycerides, triesters of glycerol. These triglycerides can be "hydrolyzed" in acid or base.

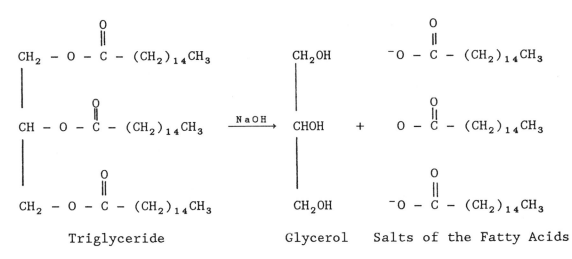

Triglyceride   Glycerol   Salts of the Fatty Acids

When the process is carried out in base, the products, the salts of the fatty acids, are soaps.

Soaps

Soap molecules can "emulsify" drops of oil, stabilizing the drops as "micelles." The non-polar "tails" of the soap molecules "dissolve" in the oil drop, while the ionic "heads" of the molecules lie on the surface of the drop. The micelles cannot coalesce into larger drops because the negatively charged micelles repel each other. The soap molecules stabilize the micelles and allow them to be washed away.

We've taken a short look at some biologically important molecules. We found the same covalent bonds in these complex molecules as we had found in simpler molecules. We found the same sorts of interactions, hydrogen bonds, dipole-dipole, etc. between molecules as we had found between simpler molecules. Now, however, the polymers are large enough that interactions between different parts of the same molecule become important in determining the shape of the polymers. The shapes of these macromolecules determine their function in biological systems.

This short introduction is meant to tantalize you into exploring more biochemistry in other courses.

23-1. Write the formula of the two possible dipeptides which can be formed from valine and glycine.

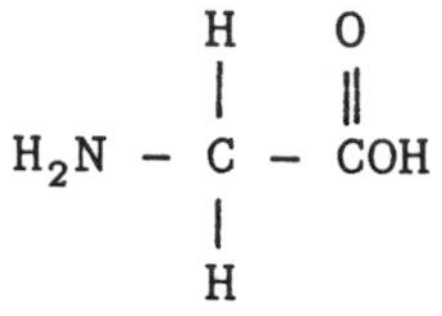

$$H_2N - \overset{\overset{\displaystyle H}{|}}{\underset{\underset{\displaystyle H}{|}}{C}} - \overset{\overset{\displaystyle O}{\|}}{C}OH$$

Glycine

and

$$H_2N - \overset{\overset{\displaystyle H}{|}}{\underset{\underset{\displaystyle CH_3 - \overset{|}{\underset{\underset{\displaystyle H}{|}}{C}} - CH_3}{|}}{C}} - \overset{\overset{\displaystyle O}{\|}}{C}OH$$

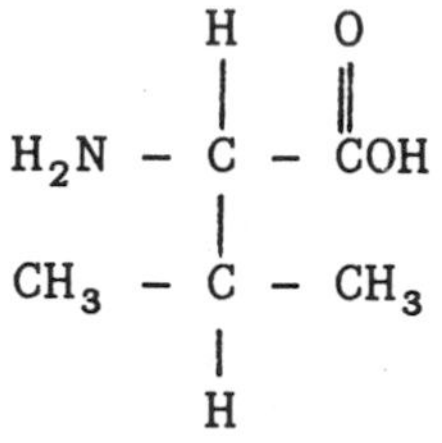

Leucine

23-2. Find the optically active (chiral) carbons, if any, in each molecule:

a. 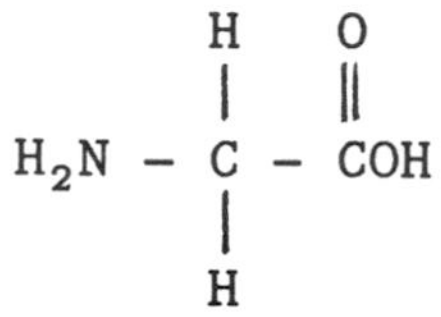 $H_2N - \overset{\overset{\displaystyle H}{|}}{\underset{\underset{\displaystyle H}{|}}{C}} - \overset{\overset{\displaystyle O}{\|}}{C}OH$

b.
$$CH_2OH$$
$$|$$
$$C = O$$
$$|$$
$$HO - C - H$$
$$|$$
$$H - C - OH$$
$$|$$
$$H - C - OH$$
$$|$$
$$CH_2OH$$

c. $H_2N - \overset{\overset{\displaystyle H}{|}}{\underset{\underset{\displaystyle CH_3 - \overset{|}{\underset{\underset{\displaystyle H}{|}}{C}} - CH_3}{|}}{C}} - \overset{\overset{\displaystyle O}{\|}}{C} - OH$

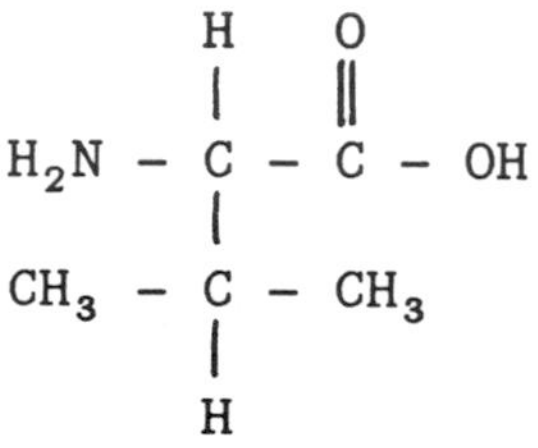

d. 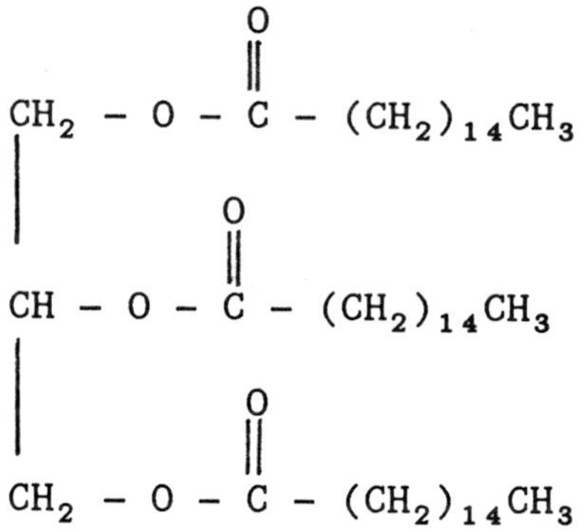
$$CH_2 - O - \overset{\overset{\displaystyle O}{\|}}{C} - (CH_2)_{14}CH_3$$
$$|$$
$$CH - O - \overset{\overset{\displaystyle O}{\|}}{C} - (CH_2)_{14}CH_3$$
$$|$$
$$CH_2 - O - \overset{\overset{\displaystyle O}{\|}}{C} - (CH_2)_{14}CH_3$$

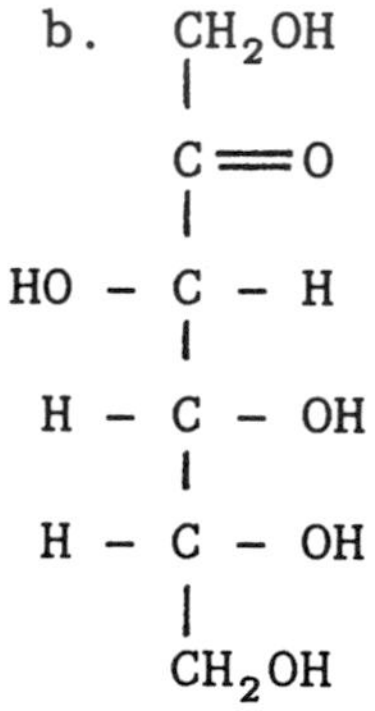

23-3. Identify which amino acid side groups can hydrogen bond with water.

a. $- CH_2CH_2SCH_3$

b. 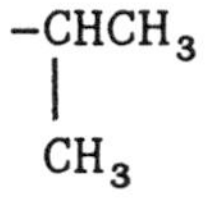 $-\overset{\overset{}{|}}{\underset{\underset{\displaystyle CH_3}{|}}{C}}HCH_3$

c. $- CH_2CH_2C\overset{\diagup\raisebox{1ex}{O}}{\diagdown\raisebox{-1ex}{OH}}$

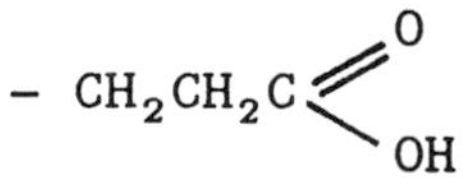

d. $-CH_2OH$

e. $-CH_2CH_2CH_2CH_2NH_2$

f. 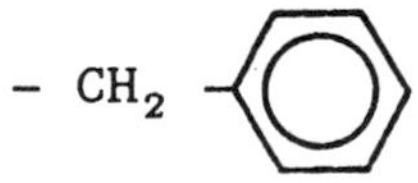 $- CH_2 -$ ⬡

23-4. Identify the amino acid side groups in 23-3 which will be charged at pH 7.

23-5. Complete the following reaction

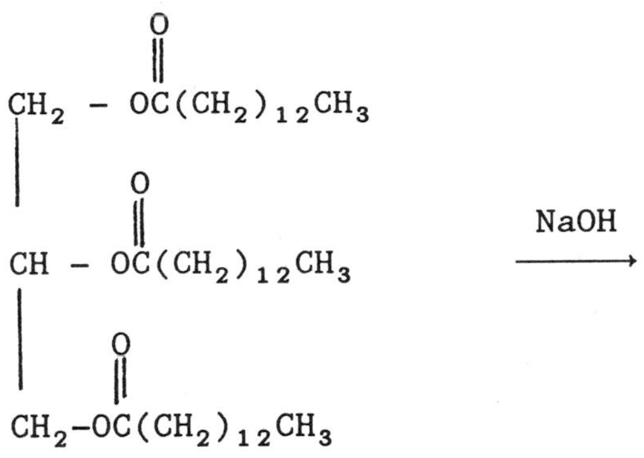

$$CH_2 - OC(CH_2)_{12}CH_3$$
(with O double bonded above the C)

$$CH - OC(CH_2)_{12}CH_3 \xrightarrow{\text{NaOH}}$$
(with O double bonded above the C)

$$CH_2 - OC(CH_2)_{12}CH_3$$
(with O double bonded above the C)

23-6. Write all the possible stereoisomers of

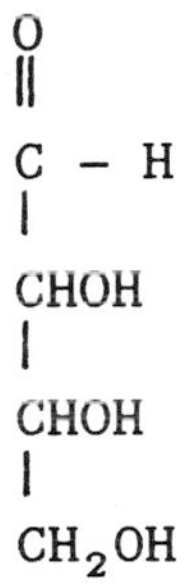

$$C - H$$
(with O double bonded above the C)
$$CHOH$$
$$CHOH$$
$$CH_2OH$$

23-1.

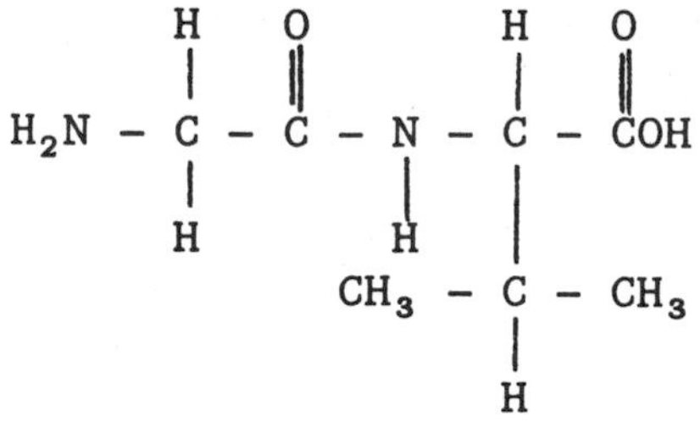

```
        H   O        H   O
        |   ||       |   ||
H₂N  -  C - C  -  N - C - COH
        |         |   |
        H         H   |
                  CH₃ - C - CH₃
                        |
                        H
```

Glycylleucine

```
          H   O           H   O
          |   ||          |   ||
and  H₂N - C - C  -  N  - C - COH
          |          |   |
          CH₃ - C - CH₃ H  H
                |
                H
```

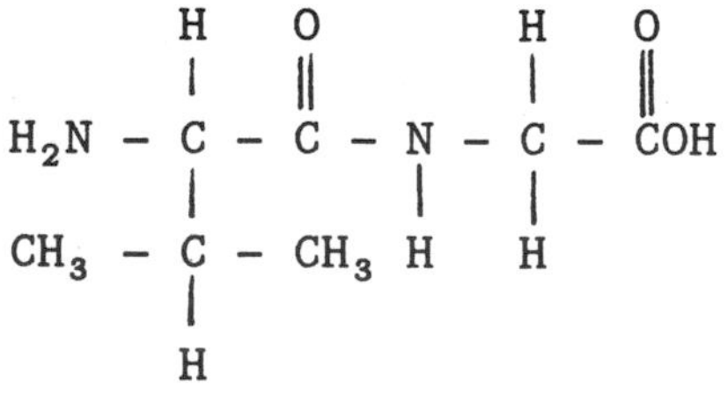

Leucylglycine

23-2.   a.   Has no chiral carbons

b. 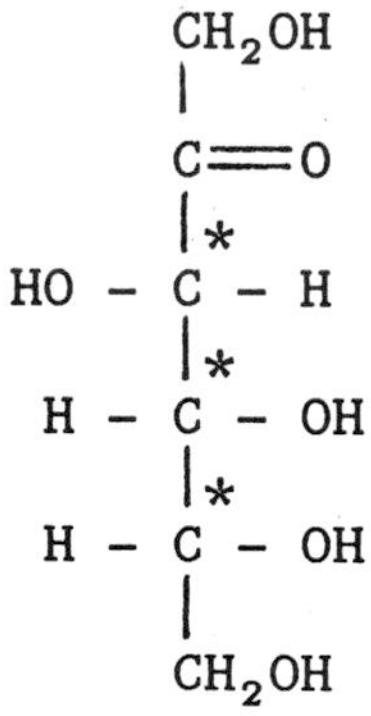
```
    CH₂OH
    |
    C══O
    |*
HO - C - H
    |*
 H - C - OH
    |*
 H - C - OH
    |
    CH₂OH
```

c.
```
        H   O
        |*  ||
H₂N  -  C - C - OH
        |
    CH₃ - C - CH₃
        |
        H
```

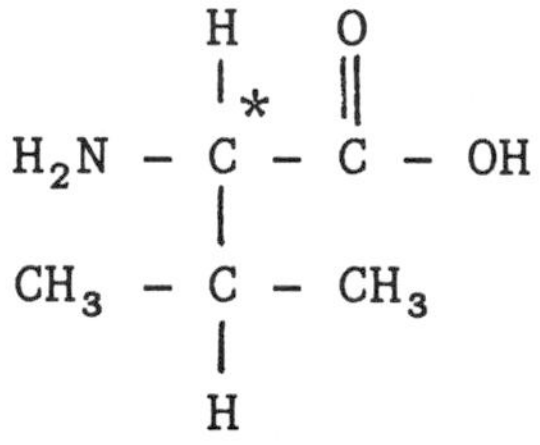

d.   Has no chiral carbons.

23-3.   c, d, e

23-4.   c, e

23-5.
```
    CH₂OH                    O
    |                        ||
→   CHOH        +    3    ⁻OC(CH₂)₁₂CH₃
    |
    CH₂OH
```

23-6.
```
        O                          O
        ||                         ||
        C - H                      C - H
        |                          |
   HO - C - H                 HO - C - H
        |                          |
   HO - C - H                  H - C - OH
        |                          |
        CH₂OH                      CH₂OH
```

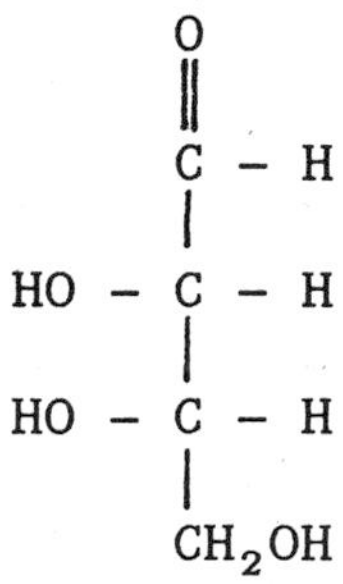

And its mirror image 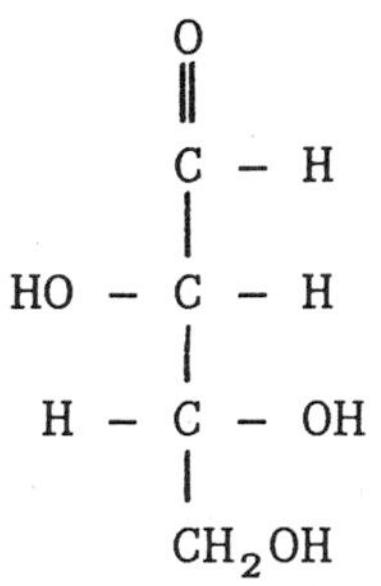   And its mirror image

CHAPTER TWENTY-FOUR: INDUSTRIAL CHEMISTRY

This chapter could be "subtitled" applied chemistry. It offers examples of the commercial applications of chemistry, the concerns of both chemists and chemical engineers.

The chemical principles you have learned in this course apply to these industrial processes, as well as the reactions you have studied in the laboratory. In this course we have tried to provide you with a foundation of chemical principles, principles you can apply in understanding chemistry in the world around you. Chemistry pervades our lives--the fertilizers and pesticides used to grow our food, the plastics, polymers and fibers used in our clothes and furniture, the additives used to preserve our food, the fuels used to provide energy, the medicines used to fight disease, the metal alloys used to construct our machines, the many compounds which pollute our air, our water and our soil. You will find chemists applying the principles of chemistry to the challenges in each of these fields.

Sometimes, a general chemistry student feels as though chemistry is a collection of different facets of science. Chemistry isn't a "linear" topic; it doesn't have a clear beginning and end. Each topic in chemistry is related to every other topic. These last three chapters--Organic, Biochemical and Industrial Chemistry--may give you a glimpse of the "whole" of chemistry, a chemist's approach to understanding the world.

24-1. Write balanced equations for the following reactions:

    a. Production of lime, CaO, by thermal decomposition of limestone, $CaCO_3$:

$$CaCO_3(s) \xrightarrow[\Delta]{}$$

    b. Oxidation of sulfur to SO_3:

$$S(s) + O_2(g) \rightarrow$$

    c. Production of $H_2SO_4(aq)$ from $SO_3(g)$ and water.

    d. Production of ammonium nitrate from ammonia and nitric acid.

    e. Reduction of iron ore, Fe_2O_3, with carbon monoxide to produce iron and carbon dioxide.

    f. The oxidation of ethanol, CH_3CH_2OH, by oxygen to produce acetic acid, CH_3COOH.

24-2. Galena (PbS) ore is "roasted" with oxygen to produce lead oxide, PbO. Lead oxide is then reduced by heating with carbon:

$$2\ PbS(s) + 3\ O_2(g) \rightarrow 2\ PbO(s) + 2\ O_2(g)$$

$$PbO(s) + C(s) \rightarrow Pb(\ell) + CO(g)$$

    a. How many <u>tons</u> of $SO_2(g)$ are produced when 1.00 <u>ton</u> of galena is processed to produce lead?

    b. How many <u>tons</u> of $CO(g)$ are produced when 1.00 <u>ton</u> of galena is processed to produce lead?

24-3. In the Hall process, aluminum is produced by electrolysis of molten aluminum oxide (Al_2O_3) mixed with NaF and AlF_3 (cryolite).

The cathode reaction is $Al^{3+} + 3\ e^- \rightarrow Al(\ell)$

The anode reaction is $2\ O^{2-} \rightarrow O_2(g) + 4\ e^-$

    a. What is the overall reaction in the electrolysis cell.

    b. 48.5 kg of Al_2O_3 are mixed with cryolite and electrolyzed. How long would a 0.800 amp current have to be passed through the mixture to completely convert all the Al^{3+} from the Al_2O_3 aluminum metal?

    c. What volume of O_2, measured at STP, would be produced in this process?

24-4. Dacron is produced by a "condensation" reaction between terephthalic acid and ethylene glycol, eliminating water.

$$HO - \overset{\overset{O}{\parallel}}{C} - \underset{}{\bigcirc} - \overset{\overset{O}{\parallel}}{C} - OH \quad + \quad H - O - CH_2 - CH_2 - O - H$$

$$+ \quad HO - \overset{\overset{O}{\parallel}}{C} - \underset{}{\bigcirc} - \overset{\overset{O}{\parallel}}{C} - OH + H - O - CH_2 - CH_2 - O - H$$

Draw the resulting polymer

24-1.   a. $CaCO_3(s) \xrightarrow{\Delta} CaO(s) + CO_2(g)$

     b. $2\ S(s) + 3\ O_2(g) \rightarrow 2\ SO_3(g)$

     c. $SO_3(g) + H_2O(\ell) \rightarrow H_2SO_4(aq)$

     d. $NH_3(g) + HNO_3(aq) \rightarrow NH_4NO_3(aq)$

     e. $Fe_2O_3(s) + 3\ CO(g) \rightarrow 2\ Fe(\ell) + 3\ CO_2(g)$

     f. $CH_3CH_2OH + O_2 \rightarrow CH_3\overset{O}{C}OH + H_2O$

24-2.   FW   PbS   239.26

     MW SO_2 64.06

     MW   CO   28.01

Therefore, 2(239.26) tons of PbS produces 2(64.06) tons of SO_2 and 2(28.01) tons of CO.

     a.   0.268 tons SO_2

     b.   0.117 tons CO

24-3.   a. $4\ Al^{3+} + 6\ O^{2-} \rightarrow 4\ Al(\ell) + 3\ O_2(g)$

     b.   476 moles $Al_2O_3 \times \dfrac{6 \text{ moles } e^-}{1 \text{ mole } Al_2O_3} \times \dfrac{96,500 \text{ C}}{\text{mole } e^-} = 2.76 \times 10^8$ C

     $2.76 \times 10^8 \text{ C} \times \dfrac{1 \text{ s}}{0.800 \text{ C}} = 3.45 \times 10^8 \text{ s} = 3.99 \times 10^3$ days

     c.   476 moles $Al_2O_3 \times \dfrac{3 \text{ moles } O_2}{2 \text{ moles } Al_2O_3} = 714$ moles O_2

     714 moles $O_2 \times \dfrac{22.4 \text{ L}}{\text{mole}} = 1.60 \times 10^4$ liters

24-4.

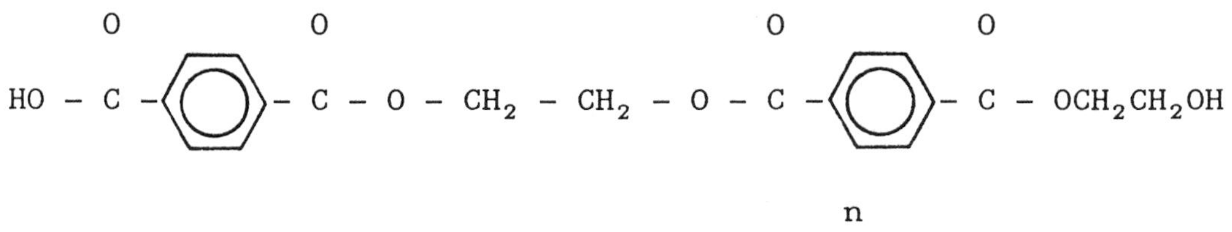

$$HO-\overset{O}{C}-\bigcirc-\overset{O}{C}-O-CH_2-CH_2-O-\overset{O}{C}-\bigcirc-\overset{O}{C}-OCH_2CH_2OH$$

$$n$$

1 2 3 4 5 6 7 8 9 0